T0250695

Lecture Notes in Computer Science

Edited by G. Goos and J. Hartmanis

455

C.A. Floudas
P.M. Pardalos

A Collection of Test Problems for Constrained Global Optimization Algorithms

Springer-Verlag

Berlin Heidelberg New York London
Paris Tokyo Hong Kong Barcelona

Authors

Christodoulos A. Floudas
Department of Chemical Engineering, Princeton University
Princeton, N.J. 08544-5263, USA

Panos M. Pardalos
Computer Science Department, The Pennsylvania State University
University Park, PA 16802, USA

CR Subject Classification (1987): G.1.6, G.1.2, F.2.1

ISBN 3-540-53032-0 Springer-Verlag Berlin Heidelberg New York
ISBN 0-387-53032-0 Springer-Verlag New York Berlin Heidelberg

Preface

A significant research activity has occurred in the area of global optimization in recent years. Many new theoretical, algorithmic, and computational contributions have resulted. Despite these numerous contributions, there still exists a lack of representative nonconvex test problems for constrained global optimization algorithms.

Test problems are of major importance for researchers interested in the algorithmic development. This book is motivated from the scarcity of global optimization test problems and represents the first systematic collection of test problems for evaluating and testing constrained global optimization algorithms. This collection includes problems arising in a variety of engineering applications, and test problems from published computational reports.

We would like to thank our students A. Aggarwal, A.R. Ciric, Chi-Geun Han, G.E. Paules IV, A.C. Kokossis, and V. Visweswaran, for their help in producing and proofreading an earlier draft of this book.

C.A. Floudas and P.M. Pardalos

June 1990

Contents

List of Tables

List of Figures

Chapter 1

Introduction

Constrained global optimization is concerned with the characterization and computation of global minima or maxima of nonconvex problems. The general constrained global minimization problem has the following form:

$$MIN_{x \in X} \qquad f(x)$$

Subject to

$$h(x) = 0$$
$$g(x) \leq 0$$

where $f(x)$ is a real valued continuous function, X is a nonempty compact set in R^n, $h(x)$ represents a set of m equality constraints and $g(x)$ represents a set of k inequality constraints.

Such problems are widespread in the mathematical modeling of real world systems for a very broad range of applications. Such applications include structural optimization, engineering design, VLSI chip design and database problems, nuclear and mechanical design, chemical engineering design and control, economies of scale, fixed charges, allocation and location problems, quadratic assignment and a number of other combinatorial optimization problems such as integer programming and related graph problems (e.g. maximum clique problem).

From the complexity point of view global optimization problems belong to the class of NP-hard problems. This means that as the input size of the problem increases the computational time required to solve the problem is expected to grow exponentially. Furthermore, even for the case of quadratic nonconvex programming it has been shown that the problem of checking if a given feasible solution is a local optimum is NP-hard ([157], [175]). In addition, the problem of determining an ε-approximate global solution remains NP-hard.

Although standard nonlinear programming algorithms will usually obtain a local minimum or a stationary point to a global optimization problem, such a local minimum will only be global when certain conditions are satisfied (such as $f(x)$ is quasi-convex and the feasible domain is convex). In general, several local minima may exist and the corresponding function values may differ substantially. The problem of designing algorithms that obtain global solutions is very difficult, since in general there is no local criterion for deciding whether a local solution is global.

Active research during the past two decades has produced a variety of deterministic and stochastic methods for determining global solutions to nonconvex nonlinear optimization problems. Reviews and books on the subject of global optimization include [59], [60], [171], [172], [80], [185], [48], [220], [152], [114]. An extensive list of references on deterministic global optimization approaches is provided in the references section of this book.

Some of the proposed algorithms for global optimization have been implemented and tested on certain problems. The efficiency of a global optimization algorithm is based upon several criteria including its effectiveness with respect to different problem classes, its speed, its capacity, and its accuracy. Since existing theory cannot itself provide measurement for these criteria, empirical computational testing is necessary. Testing and benchmarking of global optimization algorithms is a very difficult task. Three main approaches have been proposed and used to address this problem :

1. Collections of randomly generated test problems with known solution.

2. Problem instances with certain characteristics that have been used to test some aspects of specific algorithms.

3. Collection of *real-world* problems, that is, problems that model a variety of practical applications.

The first test problem generator has been proposed in [193] for convex quadratic programming. Later on, several such generators have been produced for linear and network programming problems. The generation of nontrivial test problems for global optimization programming algorithms seems to be difficult since very few papers have been devoted to that problem. Methods for automatic test problem generators have been proposed in [12], [13], [164], [159], [166], [93]. Automatic test problem generators that are available in the literature generate concave or indefinite quadratic problems, quadratic assignment problems and some classes of integer and network programming.

Some of the key features of such automatic test problem generators are the following :

1. Virtually limitless supply of test problems.

2. The ability of the user to control and know some solution characteristics.

3. The ability of the user to conduct computational experiments of parameter variation.

4. The option to generate rather store the test problems (e.g. it is enough to store the seeds of a random number generator).

In nonlinear optimization there are some excellent collections of test problems, such as the collections in [51], [57], [104], [196], [154]. Although there are many test problems in global optimization, to our knowledge there is no such a collection available in the literature. In this book a collection of test problems for constrained global optimization is provided that can be used to experiment with global optimization algorithms.

There exist different types of specific problem instances that are used to test some aspects of an algorithm. Such types of test problems include :

---

1. **Worst case test problems** : For example a global optimization test problem with an exponential number of local minima can be used to check the efficiency of an algorithm based on local searches or simulated annealing methods ([166], [172]).

2. *Standard test* problems : A test problem becomes standard if it is used frequently. Standard test problems are usually small dimension problems published in papers to illustrate the main steps of a particular algorithm. Most of the references listed at the end of this book contain standard test problems.

Up to this date, most of the reported computational experiments regarding the performance of global optimization algorithms are using either randomly generated or standard test problems. Using such test problems, reported computational experiments can be documented in a manner that allows checking and reproducing the results.

However, as George B. Dantzig said :

> *The final test of a theory is its capacity to solve the problems which originated it.*

It is therefore evident that the final testing of a global optimization algorithm is to solve problems that model practical applications. Although random and standard test problems are very useful, unfortunately it is difficult to construct them to posses characteristics that resemble problems that arise in practical applications.

This book contains many nonconvex optimization test problems that model a diverse range of practical applications. The main criteria in selecting such test problems have been (a) the size (ranging from small to medium to large), (b) the mathematical properties (exhibiting different types of nonconvexities), and (c) the degree of difficulty (resulting from the wide range of applications).

At this point it should be emphasized that it is not easy to identify what characteristics make a specific test problem difficult. It is generally agreed that the size is very important in determining the difficulty of a test problem. Dimension and density are critical factors since many global optimization algorithms take advantage of sparsity of large scale problems to reduce computational time and storage requirements. Other characteristics include the distribution of the data (e.g. symmetric versus nonsymmetric traveling salesman problem), and the number of local optima. In addition, it should be mentioned that it is not apparent whether the number of local optima is related to the complexity of the global optimization problem or the complexity of the particular algorithm. Many nonconvex problems remain NP-hard even if they have exactly one local (global) optimum.

The availability of this collection of test problems will facilitate the efforts of comparing performance and correctness of the numerous proposed global optimization algorithms. Regarding different issues on software testing and the reporting of computational results on test problems see [52], [53].

Chapters 2 and 3 include standard and randomly generated test problems for quadratic programming, and quadratically constrained problems, respectively. These problems can be used to test correctness of an algorithm. Chapter 4 presents unconstrained and constrained nonlinear programming test problems. Starting from Chapter 4, test problems are presented that arise in a variety of applications such as distillation column sequencing, blending/pooling, heat exchanger networks, phase and chemical reaction equilibrium, complex reactor networks, reactor-separator-recycle systems, mechanical design, and VLSI design. Most of these problems are difficult to solve and only the best known solution is reported.

Chapter 2

Quadratic Programming test problems

In this chapter several nonconvex quadratic programming test problems are considered. Quadratic programming has numerous applications ([172]) and plays a key role in many nonlinear programming methods. In addition, a very broad class of difficult combinatorial optimization problems such as integer programming, quadratic assignment, and the maximum clique problem can be formulated as nonconvex quadratic programming problems.

In the literature, several algorithms have been proposed and implemented for the solution of large scale concave and indefinite quadratic programming problems. Computational results are reported in [248], [162], [168], [87], and [172]. Some of the following test problems have been solved by global optimization algorithms presented in the above cited references.

2.1 Test Problem 1

2.1.1 Problem Formulation

$$MIN \qquad f(x) = c^T x - 0.5 x^T Q x$$

Subject to

$$20x_1 + 12x_2 + 11x_3 + 7x_4 + 4x_5 \leq 40$$
$$0 \leq x_i \leq 1$$

2.1.2 Data

$$c = (42, 44, 45, 47, 47.5)$$
$$Q = 100I$$

I : identity matrix

2.1.3 Global Solution

$$x^* = (1, 1, 0, 1, 0,)$$
$$f(x^*) = -17$$

2.2 Test Problem 2

2.2.1 Problem Formulation

$$MIN \qquad f(x, y) = c^T x - 0.5 x^T Q x + d^T y$$

Subject to

$$6x_1 + 3x_2 + 3x_3 + 2x_4 + x_5 \leq 6.5$$
$$10x_1 + 10x_3 + y \leq 20$$
$$0 \leq x_i \leq 1$$
$$y \geq 0$$

2.2.2 Data

$$c = (-10.5, -7.5, -3.5, -2.5, -1.5)$$
$$Q = I$$
$$d = -10$$

2.2.3 Global Solution

$$(x^*, y^*) = (0, 1, 0, 1, 1, 20)$$
$$f(x^*, y^*) = -213$$

2.3 Test Problem 3

2.3.1 Problem Formulation

$$MIN \qquad f(x, y) = c^T x - 0.5 x^T Q x + d^T y$$

Subject to

$$2x_1 + 2x_2 + y_6 + y_7 \leq 10$$
$$2x_1 + 2x_3 + y_6 + y_8 \leq 10$$
$$2x_2 + 2x_3 + y_7 + y_8 \leq 10$$
$$-8x_1 + y_6 \leq 0$$
$$-8x_2 + y_7 \leq 0$$
$$-8x_3 + y_8 \leq 0$$
$$-2x_4 - y_1 + y_6 \leq 0$$
$$-2y_2 - y_3 + y_7 \leq 0$$
$$-2y_4 - y_5 + y_8 \leq 0$$
$$0 \leq x_i \leq 1 \quad i = 1, 2, 3, 4$$
$$0 \leq y_i \leq 1 \quad i = 1, 2, 3, 4, 5$$
$$0 \leq y_9 \leq 1$$

$$0 \le y_6$$
$$0 \le y_7$$
$$0 \le y_8$$

2.3.2 Data

$$c = (5, 5, 5, 5)$$
$$Q = 10I$$
$$d = (-1, -1, -1, -1, -1, -1, -1, -1, -1)$$

2.3.3 Global Solution

$$(x^*, y^*) = (1, 1, 1, 1, 1, 1, 1, 1, 1, 3, 3, 3, 1)$$

2.4 Test Problem 4

2.4.1 Problem Formulation

$$MIN \qquad f(x, y) = 6.5x - 0.5x^2 - y_1 - 2y_2 - 3y_3 - 2y_4 - y_5$$

Subject to

$$AX \le b$$
$$0 \le X = (x, y)^T$$
$$y_i \le 1 \quad i = 3, 4$$
$$y_5 \le 2$$
$$x \in R$$
$$y \in R^5$$

2.4.2 Data

A is the following (5x6) matrix :

$$\begin{pmatrix} 1 & 2 & 8 & 1 & 3 & 5 \\ -8 & -4 & -2 & 2 & 4 & -1 \\ 2 & 0.5 & 0.2 & -3 & -1 & -4 \\ 0.2 & 2 & 0.1 & -4 & 2 & 2 \\ -0.1 & -0.5 & 2 & 5 & -5 & 3 \end{pmatrix}$$

$$b = (16, -1, 24, 12, 3)^T$$

2.4.3 Global Solution

$$(x^*, y^*) = (0, 6, 0, 1, 1, 0)$$
$$f(x^*, y^*) = -11.005$$

2.5 Test Problem 5

2.5.1 Problem Formulation

$$MIN \qquad f(x, y) = c^T x - 0.5 x^T Q x + d^T y$$

Subject to

$$AX \leq b$$
$$X = (x, y)^T$$
$$0 \leq X \leq 1$$
$$x \in R^7$$
$$y \in R^3$$

2.5.2 Data

A is the following (11x10) matrix :

$$\begin{pmatrix}
-2 & -6 & -1 & 0 & -3 & -3 & -2 & -6 & -2 & -2 \\
6 & -5 & 8 & -3 & 0 & 1 & 3 & 8 & 9 & -3 \\
-5 & 6 & 5 & 3 & 8 & -8 & 9 & 2 & 0 & -9 \\
9 & 5 & 0 & -9 & 1 & -8 & 3 & -9 & -9 & -3 \\
-8 & 7 & -4 & -5 & -9 & 1 & -7 & -1 & 3 & -2 \\
-7 & -5 & -2 & 0 & -6 & -6 & -7 & -6 & 7 & 7 \\
1 & -3 & -3 & -4 & -1 & 0 & -4 & 1 & 6 & 0 \\
1 & -2 & 6 & 9 & 0 & -7 & 9 & -9 & -6 & 4 \\
-4 & 6 & 7 & 2 & 2 & 0 & 6 & 6 & -7 & 4 \\
1 & 1 & 1 & 1 & 1 & 1 & 1 & 1 & 1 & 1 \\
-1 & -1 & -1 & -1 & -1 & -1 & -1 & -1 & -1 & -1
\end{pmatrix}$$

$$b = (-4, 22, -6, -23, -12, -3, 1, 12, 15, 9, -1)^T$$
$$Q = 10I$$
$$d = (10, 10, 10)$$
$$c = (-20, -80, -20, -50, -60, -90, 0)$$

2.5.3 Global Solution

$$(x^*, y^*) = (1, 0.907, 0, 1, 0.715, 1, 0, 0.917, 1, 1)$$
$$f(x^*, y^*) = 771.985$$

2.6 Test Problem 6

2.6.1 Problem Formulation

$$MIN \qquad f(x, y) = c^T x - 0.5 x^T Q x$$

Subject to

$$Ax \leq b$$
$$0 \leq x_i \leq 1 \quad i = 1, 2, ..., 10$$
$$x \in R^{10}$$

2.6.2 Data

A is the following (5x10) matrix :

$$\begin{pmatrix} -2 & -6 & -1 & 0 & -3 & -3 & -2 & -6 & -2 & -2 \\ 6 & -5 & 8 & -3 & 0 & 1 & 3 & 8 & 9 & -3 \\ -5 & 6 & 5 & 3 & 8 & -8 & 9 & 2 & 0 & -9 \\ 9 & 5 & 0 & -9 & 1 & -8 & 3 & -9 & -9 & -3 \\ -8 & 7 & -4 & -5 & -9 & 1 & -7 & -1 & 3 & -2 \end{pmatrix}$$

$$b = (-4, 22, -6, -23, -12)^T$$
$$Q = 100I$$
$$c = (48, 42, 48, 45, 44, 41, 47, 42, 45, 46)$$

2.6.3 Global Solution

$$x^* = (1, 0, 0, 1, 1, 1, 0, 1, 1, 1)$$
$$f(x^*) = -39$$

2.7 Test Problem 7

2.7.1 Problem Formulation

Let the polytope P be defined by :

$$P = \{x : Ax \leq b, \ x \geq 0\} \subseteq R^{20}$$

where A is a (10x20) matrix and $b \in R^{20}$. The A^T matrix and the vector b are :

$$
\begin{pmatrix}
-3 & 7 & 0 & -5 & 1 & 1 & 0 & 2 & -1 & 1 \\
7 & 0 & -5 & 1 & 1 & 0 & 2 & -1 & -1 & 1 \\
0 & -5 & 1 & 1 & 0 & 2 & -1 & -1 & -9 & 1 \\
-5 & 1 & 1 & 0 & 2 & -1 & -1 & -9 & 3 & 1 \\
1 & 1 & 0 & 2 & -1 & -1 & -9 & 3 & 5 & 1 \\
1 & 0 & 2 & -1 & -1 & -9 & 3 & 5 & 0 & 1 \\
0 & 2 & -1 & -1 & -9 & 3 & 5 & 0 & 0 & 1 \\
2 & -1 & -1 & -9 & 3 & 5 & 0 & 0 & 1 & 1 \\
-1 & -1 & -9 & 3 & 5 & 0 & 0 & 1 & 7 & 1 \\
-1 & -9 & 3 & 5 & 0 & 0 & 1 & 7 & -7 & 1 \\
-9 & 3 & 5 & 0 & 0 & 1 & 7 & -7 & -4 & 1 \\
3 & 5 & 0 & 0 & 1 & 7 & -7 & -4 & -6 & 1 \\
5 & 0 & 0 & 1 & 7 & -7 & -4 & -6 & -3 & 1 \\
0 & 0 & 1 & 7 & -7 & -4 & -6 & -3 & 7 & 1 \\
0 & 1 & 7 & -7 & -4 & -6 & -3 & 7 & 0 & 1 \\
1 & 7 & -7 & -4 & -6 & -3 & 7 & 0 & -5 & 1 \\
7 & -7 & -4 & -6 & -3 & 7 & 0 & -5 & 1 & 1 \\
-7 & -4 & -6 & -3 & 7 & 0 & -5 & 1 & 1 & 1 \\
-4 & -6 & -3 & 7 & 0 & -5 & 1 & 1 & 0 & 1 \\
-6 & -3 & 7 & 0 & -5 & 1 & 1 & 0 & 2 & 1
\end{pmatrix}
$$

$$ b = (-5, 2, -1, -3, 5, 4, -1, 0, 9, 40)^T $$

Consider the separable concave quadratic programming problem given by :

$$ MIN \qquad f(x) = -0.5 \sum_{i=1}^{20} \lambda_i (x_i - \alpha_i)^2 $$

Subject to

$$ x \in P $$

where $\lambda = (\lambda_1, ..., \lambda_{20}) \geq 0$ is the set of eigenvalues and the vector $\alpha = (\alpha_1, ..., \alpha_{20})$ is the unconstrained maximum of $f(x)$.

2.7.2   Data

The following test problems are of the form described above (fixed P) with different values of λ and α.

Case 1 :

$$f(x) = -0.5 \sum_{i=1}^{20} (x_i - 2)^2$$

Best Known Solution

$x^* = (0, 0, 28.8024, 0, 0, 4.1792, 0, 0, 0, 0, 0, 0, 0, 0, 0.6188, 4.0933, 0, 2.3064, 0, 0)$
$f(x^*) = -394.7506$

Case 2 :

$$f(x) = -0.5 \sum_{i=1}^{20} (x_i + 5)^2$$

Best Known Solution

$x^* = (0, 0, 28.8024, 0, 0, 4.1792, 0, 0, 0, 0, 0, 0, 0, 0, 0.6187, 4.0933, 0, 2.3064, 0, 0)$
$f(x^*) = -884.75058$

Case 3 :

$$f(x) = -10 \sum_{i=1}^{20} (x_i)^2$$

Best Known Solution

$x^* = (0, 0, 28.8024, 0, 0, 4.1792, 0, 0, 0, 0, 0, 0, 0, 0, 0.6187, 4.0933, 0, 2.3064, 0, 0)$
$f(x^*) = -8695.01193$

Case 4 :

$$f(x) = -0.5 \sum_{i=1}^{20} (x_i - 8)^2$$

Best Known Solution

$$x^* = (0, 0, 28.8024, 0, 0, 4.1792, 0, 0, 0, 0, 0, 0, 0, 0, 0.6187, 4.0933, 0, 2.3064, 0, 0)$$
$$f(x^*) = -754.75062$$

Case 5 :

$$f(x) = -0.5 \sum_{i=1}^{20} i(x_i - 2)^2$$

Best Known Solution

$$x^* = (0, 0, 0, 0.9949, 0, 0, 0, 0, 0, 0, 0, 0.9299, 0, 0, 0, 7.4117, 0, 12.6736, 0, 17.9899)$$
$$f(x^*) = -4105.2779$$

Extensive computational experiments ([162], [87], [248], [172]) with concave quadratic problems of the above form suggests that on the average, the difficulty of the problem depends on the size of the eigenvalues (the curvature of the objective function) and the location of the unconstrained maximum α of $f(x)$. It has been observed that if α belongs to the interior of P, then such problems are the most difficult to solve. On the other hand, if α lies outside of P, then the corresponding test problems are, on the average, easier.

2.8   Test Problem 8

An important class of global optimization problems is the minimum concave cost transportation problems. A recent survey regarding complexity, algorithms, and applications of minimum concave cost network problems is [92]. The general formulation is given as follows :

2.8.1 Problem Formulation

$$MIN \qquad f(x) = \sum_{i=1}^{m}\sum_{j=1}^{n}(c_{ij}x_{ij} + d_{ij}x_{ij}^2)$$

Subject to

$$\sum_{i=1}^{m} x_{ij} = b_j \quad j = 1,...,n$$

$$\sum_{j=1}^{n} x_{ij} = a_i \quad i = 1,...,m$$

$$x_{ij} \geq 0$$

where

$$d_{ij} \leq 0, \quad \sum_{i=1}^{m} a_i = \sum_{j=1}^{n} b_j$$

This problem features $n + m$ equality constraints and nm variables. There is exactly one redundant equality constraint. When any one of the constraints is dropped the remaining form a linearly independent system of constraints. Hence, a basic vector for the transportation problem consists of $n + m - 1$ basic variables. In addition, if all a_i, b_j are positive integers, then every basic solution is an integer solution.

2.8.2 Data

$$n = 4, m = 6$$

$$a = (8, 24, 20, 24, 16, 12)^T$$

$$b = (29, 41, 13, 21)^T$$

The $C = (c_{ij})$ matrix is :

$$\begin{pmatrix} 300 & 270 & 460 & 800 \\ 740 & 600 & 540 & 380 \\ 300 & 490 & 380 & 760 \\ 430 & 250 & 390 & 600 \\ 210 & 830 & 470 & 680 \\ 360 & 290 & 400 & 310 \end{pmatrix}$$

The $D = (d_{ij})$ matrix is :

$$\begin{pmatrix} -7 & -4 & -6 & -8 \\ -12 & -9 & -14 & -7 \\ -13 & -12 & -8 & -4 \\ -7 & -9 & -16 & -8 \\ -4 & -10 & -21 & -13 \\ -17 & -9 & -8 & -4 \end{pmatrix}$$

2.8.3 Global Solution

$$x_{11}^* = 6, x_{12}^* = 2, x_{22}^* = 3, x_{24}^* = 21,$$
$$x_{31}^* = 20, x_{41}^* = 24, x_{51}^* = 3, x_{53}^* = 13,$$
$$x_{62}^* = 12$$
$$f(x^*) = 15639$$

2.9 Test Problem 9

Given an undirected graph $G(V, E)$ where V is a set of vertices and E is a set of edges, a clique is defined to be a set of vertices that is completely interconnected. The maximum clique problem consists of determining a clique of maximum cardinality.

The maximum clique problem can be stated ([169]) as a nonconvex quadratic programming problem over the unit simplex, as follows :

$$MAX \qquad f(x) = \sum_{(i,j) \in E} x_i x_j$$

Subject to

$$\sum_{v_i \in V} x_i = 1$$

$$x_i \geq 0 \quad i = 1, ..., |V|$$

If k is the size of the maximum clique of G, then $f(x^*) = 0.5(1 - (1/k))$. The global maximum x^* is defined by $x_i^* = (1/k)$, if the vertex v_i belongs to the maximum clique, and zero otherwise.

2.9.1 Problem Formulation

$$MAX \qquad f(x) = \sum_{i=1}^{9} x_i x_{i+1} + \sum_{i=1}^{8} x_i x_{i+2} + x_1 x_9 + x_1 x_{10} + x_2 x_{10} + x_1 x_5 + x_4 x_7$$

Subject to

$$\sum_{i=1}^{10} x_i = 1$$

$$x_i \geq 0 \quad i = 1, ..., 10$$

2.9.2 Global Solution

$$x^* = (0, 0, 0, 0.25, 0.25, 0.25, 0.25, 0, 0, 0)$$

$$f(x^*) = 0.375$$

2.10 Test Problem 10

2.10.1 Problem Formulation

$$MIN \qquad f(X, Y) = -0.5 \sum_{1}^{10} \lambda_i (x_i - \alpha_i)^2 + 0.5 \sum_{1}^{10} \mu_i (x_i - \beta_i)^2$$

Subject to

$$A_1 X + A_2 Y \leq b$$
$$X, Y \geq 0$$
$$X \in R^{10}$$
$$Y \in R^{10}$$

The first summation represents the concave part of the objective function and the second summation the convex part.

2.10.2 Data

$$\lambda = (63, 15, 44, 91, 45, 50, 89, 58, 86, 82)$$
$$\mu = (42, 98, 48, 91, 11, 63, 61, 61, 38, 26)$$
$$\alpha = (-19, -27, -23, -53, -42, 26, -33, -23, 41, 19)$$
$$\beta = (-52, -3, 81, 30, -85, 68, 27, -81, 97, -73)$$

A_1 is the following (10x10) matrix corresponding to the concave variables :

$$
\begin{pmatrix}
3 & 5 & 5 & 6 & 4 & 4 & 5 & 6 & 4 & 4 \\
5 & 4 & 5 & 4 & 1 & 4 & 4 & 2 & 5 & 2 \\
1 & 5 & 2 & 4 & 7 & 3 & 1 & 5 & 7 & 6 \\
3 & 2 & 6 & 3 & 2 & 1 & 6 & 1 & 7 & 3 \\
6 & 6 & 6 & 4 & 5 & 2 & 2 & 4 & 3 & 2 \\
5 & 5 & 2 & 1 & 3 & 5 & 5 & 7 & 4 & 3 \\
3 & 6 & 6 & 3 & 1 & 6 & 1 & 6 & 7 & 1 \\
1 & 2 & 1 & 7 & 8 & 7 & 6 & 5 & 8 & 7 \\
8 & 5 & 2 & 5 & 3 & 8 & 1 & 3 & 3 & 5 \\
1 & 1 & 1 & 1 & 1 & 1 & 1 & 1 & 1 & 1
\end{pmatrix}
$$

A_2 is the following (10x10) matrix corresponding to the convex variables :

$$
\begin{pmatrix}
8 & 2 & 4 & 1 & 1 & 1 & 2 & 1 & 7 & 3 \\
3 & 6 & 1 & 7 & 7 & 5 & 8 & 7 & 2 & 1 \\
1 & 7 & 2 & 4 & 7 & 5 & 3 & 4 & 1 & 2 \\
7 & 7 & 8 & 2 & 3 & 4 & 5 & 8 & 1 & 2 \\
7 & 5 & 3 & 6 & 7 & 5 & 8 & 4 & 6 & 3 \\
4 & 1 & 7 & 3 & 8 & 3 & 1 & 6 & 2 & 8 \\
4 & 3 & 1 & 4 & 3 & 6 & 4 & 6 & 5 & 4 \\
2 & 3 & 5 & 5 & 4 & 5 & 4 & 2 & 2 & 8 \\
4 & 5 & 5 & 6 & 1 & 7 & 1 & 2 & 2 & 4 \\
1 & 1 & 1 & 1 & 1 & 1 & 1 & 1 & 1 & 1
\end{pmatrix}
$$

$$b = (380, 415, 385, 405, 470, 415, 400, 460, 400, 200)^T$$

2.10.3 Best Known Solution

$$X^* = (0, 0, 0, 0, 0, 4.348, 0, 0, 0, 0,)$$
$$Y^* = (0, 0, 0, 62.609, 0, 0, 0, 0, 0, 0,)$$
$$f(X^*, Y^*) = 49318$$

Chapter 3

Quadratically Constrained test problems

Very few global optimization algorithms have been proposed for nonconvex quadratically constrained problems [6]. The general problem is intrinsically difficult on the grounds of the variety of potential nonconvexities that may occur. Separable quadratic constraints, complementarity type constraints or integer constraints may be involved. Notice that every simple integer constraint $x \in \{0,1\}$ can be written as :

$$-x^2 + x \leq 0, \ 0 \leq x \leq 1$$

or

$$x^2 - x = 0.$$

Next, a few standard quadratically constrained problems are presented.

3.1   Test Problem 1

This test problem is taken from [104] (Problem 106). It features a linear objective function subject to 6 inequality constraints (out of which 3 are nonconvex). For the bounds on the eight variables, there are 16 additional inequality constraints.

3.1.1 Problem Formulation

$$MIN \quad x_1 + x_2 + x_3$$

$$s.t. \quad -1 + 0.0025(x_4 + x_6) \leq 0$$
$$-1 + 0.0025(-x_4 + x_5 + x_7) \leq 0$$
$$-1 + 0.01(-x_5 + x_8) \leq 0$$
$$100x_1 - x_1x_6 + 833.33252x_4 - 83333.333 \leq 0$$
$$x_2x_4 - x_2x_7 - 1250x_4 + 1250x_5 \leq 0$$
$$x_3x_5 - x_3x_8 - 2500x_5 + 1250000 \leq 0$$

$$100 \leq x_1 \leq 10000$$
$$1000 \leq x_2 \leq 10000$$
$$1000 \leq x_3 \leq 10000$$
$$10 \leq x_4 \leq 1000$$
$$10 \leq x_5 \leq 1000$$
$$10 \leq x_6 \leq 1000$$
$$10 \leq x_7 \leq 1000$$
$$10 \leq x_8 \leq 1000$$

3.1.2 Best Known Solution

$$x^* = (579.31, 1359.97, 5109.97, 182.02, 295.6, 217.98, 286.42, 395.60)$$
$$f(x^*) = 7049.25$$

3.2 Test Problem 2

The second example consists of a nonconvex quadratic objective function subject to 6 inequality constraints all of which are nonconvex quadratic. There are 10 inequality constraints representing the bounds on the five variables. This test problem is taken from Colville's collection [51].

3.2.1 Problem Formulation

$$MIN \qquad 37.293239x_1 + 0.8356891x_1x_5 + 5.3578547x_3^2 - 40792.141$$

$s.t.$
$$- 0.0022053x_3x_5 + 0.0056858x_2x_5 + 0.0006262x_1x_4 - 6.665593 \leq 0$$
$$0.0022053x_3x_5 - 0.0056858x_2x_5 - 0.0006262x_1x_4 - 85.334407 \leq 0$$
$$0.0071317x_2x_5 + 0.0021813x_3^2 + 0.0029955x_1x_2 - 29.48751 \leq 0$$
$$-0.0071317x_2x_5 - 0.0021813x_3^2 - 0.0029955x_1x_2 + 9.48751 \leq 0$$
$$0.0047026x_3x_5 + 0.0019085x_3x_4 + 0.0012547x_1x_3 - 15.699039 \leq 0$$
$$-0.0047026x_3x_5 - 0.0019085x_3x_4 - 0.0012547x_1x_3 + 10.699039 \leq 0$$
$$78 \leq x_1 \leq 102$$
$$33 \leq x_2 \leq 45$$
$$27 \leq x_3 \leq 45$$
$$27 \leq x_4 \leq 45$$
$$27 \leq x_5 \leq 45$$

3.2.2 Best Known Solution

$$x^* = (78, 33, 29.9953, 45, 36.7758)$$
$$f(x^*) = -30665.5387$$

3.3 Test Problem 3

This test problem is taken from [100]. It involves a minimization of a concave quadratic function subject to linear and quadratic constraints. There exist 18 local minima. This problem can be easily decomposed into three smaller independent problems.

3.3.1 Problem Formulation

min $\qquad f(x) = -25(x_1-2)^2 - (x_2-2)^2 - (x_3-1)^2 - (x_4-4)^2 - (x_5-1)^2 - (x_6-4)^2$

$$s.t. \quad (x_3 - 3)^2 + x_4 \geq 4$$
$$(x_5 - 3)^2 + x_6 \geq 4$$
$$x_1 - 3x_2 \leq 2$$
$$-x_1 + x_2 \leq 2$$
$$x_1 + x_2 \leq 6$$
$$x_1 + x_2 \geq 2$$
$$1 \leq x_3 \leq 5$$
$$0 \leq x_4 \leq 6$$
$$1 \leq x_5 \leq 5$$
$$0 \leq x_6 \leq 10$$
$$0 \leq x_1$$
$$0 \leq x_2$$

3.3.2 Global Solution

$$x^* = (5, 1, 5, 0, 5, 10)$$
$$f(x^*) = -310$$

3.4 Test Problem 4

This test problem is taken from [30]. It involves a minimization of a linear function subject to a set of linear constraints and one reverse convex constraint.

3.4.1 Problem Formulation

$$\min \quad f(x) = -2x_1 + x_2 - x_3$$

$$s.t. \quad x_1 + x_2 + x_3 \leq 4$$
$$x_1 \leq 2$$
$$x_3 \leq 3$$

$$3x_2 + x_3 \leq 6$$
$$x_1, x_2, x_3 \geq 0$$
$$x^T B^T Bx - 2r^T Bx + \| r \|^2 - 0.25 \| b - v \|^2 \geq 0$$

where B is the following (3x3)matrix :

$$\begin{pmatrix} 0 & 0 & 1 \\ 0 & -1 & 0 \\ -2 & 1 & -1 \end{pmatrix}$$

$$b = (3, 0, -4)$$
$$v = (0, -1, -6)$$
$$r = (1.5, -0.5, -5)$$

3.4.2 Global Solution

$$x^* = (0.5, 0, 3)$$
$$f(x^*) = -4$$

A procedure for constructing reverse convex test problems with known solution is described in [30] and [31]. In addition, several medium size example problems are given.

Chapter 4

Nonlinear Programming test problems

In this chapter unconstrained and constrained general nonconvex nonlinear programming test problems are presented. The unconstrained test problems are of known global solution, while most of the constrained nonlinear programming problems feature either separable concave terms (non-quadratic) in the objective function or involve polynomial constraints, and only the best known solution is reported.

4.1 Test Problem 1

This problem is taken from [244]. It features the unconstrained minimization of a nonconvex polynomial function in one variable.

4.1.1 Problem Formulation

$$\text{MIN } x^6 - \frac{52}{25}x^5 + \frac{39}{80}x^4 + \frac{71}{10}x^3 - \frac{79}{20}x^2 - x + \frac{1}{10}$$

$$-2 \le x \le 11$$

4.1.2 Problem Statistics

Number of Variables	1
Number of Linear Constraints	0
Number of Nonlinear Constraints	0

4.1.3 Global Solution

- Objective value = -29763.233

- Variable : $x^* = 10$.

4.2 Test Problem 2 : Goldstein and Price's Function

This problem is taken from [88]. It features the unconstrained minimization of a nonconvex function in two variables.

4.2.1 Problem Formulation

$$\text{MIN } f(x, y)$$

where

$$f(x, y) = [1 + (x + y + 1)^2(19 - 14x + 3x^2 - 14y + 6xy + 3y^2)]$$
$$\cdot [30 + (2x - 3y)^2(18 - 32x + 12x^2 + 48y - 36xy + 27y^2)]$$

4.2.2 Problem Statistics

Number of Variables	2
Number of Linear Constraints	0
Number of Nonlinear Constraints	0

4.2.3 Global Solution

- Objective value = 3.0

- Variables : $x^* = 0$, $y^* = -1$.

4.3   Test Problem 3

This problem is taken from [211]. It features a nonconvex objective function subject to three linear constraints. The bounds on the variables introduce six linear inequality constraints.

4.3.1   Problem Formulation

$$\text{MIN } x_1^{0.6} + x_2^{0.6} - 6x_1 - 4u_1 + 3u_2$$

$$
\begin{aligned}
s.t. \quad x_2 - 3x_1 - 3u_1 &= 0 \\
x_1 + 2u_1 &\leq 4 \\
x_2 + 2u_2 &\leq 4 \\
x_1 &\leq 3 \\
u_2 &\leq 1 \\
x_1, x_2, u_1, u_2 &\geq 0
\end{aligned}
$$

4.3.2   Problem Statistics

Number of Variables	4
Number of Linear Constraints	3
Number of Nonlinear Constraints	0

4.3.3   Best Known Solution

- Objective value = - 4.5142

- Variables : $x^* = (4/3, 4)$, $u^* = (0, 0)$.

4.4   Test Problem 4

This problem is also taken from [211]. It features a nonconvex objective function subject to three linear constraints. The bounds on the variables introduce six linear inequality constraints.

4.4.1 Problem Formulation

$$\text{MIN} \ x_1^{0.6} + 2x_2^{0.6} + 2u_1 - 2x_2 - u_2$$

$$
\begin{aligned}
s.t. \quad x_2 - 3x_1 - 3 &= 0 \\
x_1 + 2u_1 &\leq 4 \\
x_2 + u_2 &\leq 4 \\
x_1 &\leq 3 \\
u_2 &\leq 2 \\
x_1, x_2, u_1, u_2 &\geq 0
\end{aligned}
$$

4.4.2 Problem Statistics

Number of Variables	4
Number of Linear Constraints	3
Number of Nonlinear Constraints	0

4.4.3 Best Known Solution

- Objective value = - 2.07

- Variables : $x^* = (4/3, 4)$, $u^* = (0, 0)$.

4.5 Test Problem 5

This problem is taken from [211]. It features a nonconvex objective function subject to six linear constraints. The bounds on the variables introduce nine linear inequality constraints.

4.5.1 Problem Formulation

$$\text{MIN} \ x_1^{0.6} + x_2^{0.6} + x_3^{0.4} + 2u_1 + 5u_2 - 4x_3 - u_3$$

$$s.t. \quad x_2 - 3x_1 - 3u_1 = 0$$
$$x_3 - 2x_2 - 2u_2 = 0$$
$$4u_1 - u_3 = 0$$
$$x_1 + 2u_1 \leq 4$$
$$x_2 + u_2 \leq 4$$
$$x_3 + u_3 \leq 6$$
$$x_1 \leq 3$$
$$u_2 \leq 2$$
$$x_3 \leq 4$$
$$x_1, x_2, x_3, u_1, u_2, u_3 \geq 0$$

4.5.2 Problem Statistics

Number of Variables	6
Number of Linear Constraints	6
Number of Nonlinear Constraints	0

4.5.3 Global solution

- Objective value = - 11.96

- Variables : $x^* = (0.67, 2, 4)$, $u^* = (0, 0, 0)$.

4.6 Test Problem 6

This problem features a linear objective function subject to two nonlinear inequality polynomial constraints. The bounds on the two variables introduce four additional inequality constraints. The feasible region is almost disconnected.

4.6.1 Problem Formulation

$$\text{MIN} \quad - x - y$$
$$y \leq 2x^4 - 8x^3 + 8x^2 + 2$$

$$y \leq 4x^4 - 32x^3 + 88x^2 - 96x + 36$$

$$0 \leq x \leq 3$$

$$0 \leq y \leq 4$$

4.6.2 Problem Statistics

Number of Variables	2
Number of Linear Constraints	0
Number of Nonlinear Constraints	2

4.6.3 Best Known Solution

- Objective value = -5.5079

- Variables : $x^* = 2.3295$, $y^* = 3.1783$.

4.7 Test Problem 7

This problem is taken from [208]. It features a convex objective function subject to a nonlinear equality constraint. The bounds on the variables introduce four additional linear inequality constraints.

4.7.1 Problem Formulation

$$\text{MIN} \quad -12x - 7y + y^2$$

$$-2x^4 + 2 - y = 0$$

$$0 \leq x \leq 2$$

$$0 \leq y \leq 3$$

4.7.2 Problem Statistics

Number of Variables	2
Number of Linear Constraints	0
Number of Nonlinear Constraints	1

4.7.3 Best Known Solution

- Objective value = -16.73889

- Variables : $x^* = 0.71751$, $y^* = 1.470$.

4.8 Test Problem 8

This problem is taken from [241]. It features the optimal design of a heat exchanger network with one cold stream and two hot streams. The objective function is nonlinear, and there are four linear and four nonlinear equality constraints. The bounds on the temperatures introduce four additional linear inequality constraints.

4.8.1 Problem Formulation

$$\text{MIN} \quad \Phi = 35A_1^{0.6} + 35A_2^{0.6}$$

$$
\begin{aligned}
s.t. \quad A_1 &= \frac{Q_1}{U_1 \Delta T_{lm}^1} \\
A_2 &= \frac{Q_2}{U_2 \Delta T_{lm}^2} \\
Q_1 &= F^C C_p (T_3 - TC^{in}) \\
Q_2 &= F^C C_p (TC^{out} - T_3) \\
Q_1 &= F_1^H C_p (TH_1^{in} - T1) \\
Q_2 &= F_2^H C_p (TH_2^{in} - T2) \\
\Delta T_{lm}^1 &= \frac{(T_1 - TC^{in}) - (TH_1^{in} - T_3)}{\ln\left(\frac{T_1 - TC^{in}}{TH_1^{in} - T_3}\right)} \\
\Delta T_{lm}^2 &= \frac{(T_2 - T3) - (TH_2^{in} - TC^{out})}{\ln\left(\frac{T_2 - T3}{TH_2^{in} - TC^{out}}\right)} \\
TC^{in} &\leq T_3 \leq TC^{out} \\
T_1 &\leq TH_1^{in}, \quad T_2 \leq TH_2^{in}
\end{aligned}
$$

The problem variables are T_1, T_2, T_3, Q_1, Q_2, A_1 and A_2.

4.8.2 Data

- Inlet Temperatures : $TC_{in} = 100^0$, $TH_1^{in} = 600^o$, $TH_2^{in} = 900^o$

- Outlet Temperature : $TC^{out} = 300^o$

- Flow Rates : $F^C = F_1^H = F_2^H = 10,000.$

- Specific Heat of Streams : $C_p = 1$

- Heat Transfer Coefficients : $U_1 = U_2 = 200.$

4.8.3   Problem Statistics

Number of Variables	7
Number of Linear Constraints	4
Number of Nonlinear Constraints	4

4.8.4   Best Known Solution

- Objective value = 189.3

- Temperatures : $T_1^* = 600$, $T_2^* = 700$, $T_3^* = 100.$

4.9   Test Problem 9

This problem is taken from [241]. It features a nonlinear objective function, six linear equality constraints and three nonlinear equality constraints. The bounds on the temperatures introduce seven additional linear inequality constraints.

4.9.1 Problem Formulation

$$\text{MIN} \quad \Phi = (\frac{Q_1}{U_1 \Delta T_{lm}^1})^{0.6} + (\frac{Q_2}{U_2 \Delta T_{lm}^2})^{0.6} + (\frac{Q_3}{U_3 \Delta T_{lm}^3})^{0.6}$$

$$s.t. \quad Q_1 = FC_p(T_1 - T^{in})$$
$$Q_2 = FC_p(T_2 - T_1)$$
$$Q_3 = FC_p(T^{out} - T_2)$$
$$Q_1 = FC_p(t_1^{in} - t1)$$
$$Q_2 = FC_p(t_2^{in} - t2)$$
$$Q_3 = FC_p(t_3^{in} - t3)$$
$$\Delta T_{lm}^1 = \frac{(t_1 - T^{in}) - (t_1^{in} - T_1)}{\ln\left(\frac{t_1 - T^{in}}{t_1^{in} - T_1}\right)}$$
$$\Delta T_{lm}^2 = \frac{(t_2 - T_1) - (t_2^{in} - T_2)}{\ln\left(\frac{t_2 - T_1}{t_2^{in} - T_2}\right)}$$
$$\Delta T_{lm}^3 = \frac{(t_3 - T_2) - (t_3^{in} - T^{out})}{\ln\left(\frac{t_3 - T_2}{t_3^{in} - T^{out}}\right)}$$
$$T^{in} \leq T_1, T_2 \leq T^{out}$$
$$t_1 \leq t_1^{in}, \quad t_2 \leq t_2^{in}, \quad t_3 \leq t_3^{in}$$

The problem variables are T_1, T_2, t_1, t_2, t_3, Q_1, Q_2, and Q_3.

4.9.2 Data

- Inlet Temperatures : $T_{in} = 100°$, $t_1^{in} = 300°$, $t_2^{in} = 400°$, $t_3^{in} = 600°$

- Cold Stream Outlet Temperature : $T^{out} = 500°$

- Heat Capacity of Streams : $FC_p = 10^5$

- Heat Transfer Coefficients : $U_1 = 120$, $U_2 = 80$, $U_3 = 40$

4.9.3 Problem Statistics

Number of Variables	11
Number of Linear Constraints	6
Number of Nonlinear Constraints	3

4.9.4 Best Known Solution

- Objective value = 7049

- Intermediate Cold Stream Temperatures : $T_1^* = 181.9^o$, $T_2^* = 295^o$

- Hot Stream Outlet Temperatures : $t_1^* = 281.1^o$, $t_2^* = 286.4^o$, $t_3^* = 395.5^o$.

Chapter 5

Distillation Column Sequencing test problems

Starting from this chapter several test problems that arise in a variety of engineering applications are considered. The problems vary in the degree of difficulty and only the best known solution is reported.

5.1 Problem Statement

Separation processes constitute a significant portion of the total capital investment and operating expense for a chemical plant and a lot of interest has been generated in the development of systematic approaches that select an optimal sequence of separation columns. The problem can be stated as :

> *A single multicomponent feed stream of known conditions (i.e. flowrate, composition, temperature and pressure) is given which has to be separated into a number of multicomponent products of specified compositions. The problem is then to synthesize an optimal distillation sequence, allowing the use of nonsharp separators, that separates the single multicomponent feed into several multicomponent products and satisfies the criterion of* minimum total annual cost.

Recent work on separation sequencing based upon optimization approaches addressed the general separation problem with sharp columns ([72], [239], [75], [78]), and the nonsharp distillation sequencing problem ([240], [4]). The mathematical formulation of the nonsharp distillation problem for a given set of columns corresponds to a non-convex NLP model [4] that involves a nonlinear objective function (total annual cost) subject to a linear and nonlinear set of constraints (total and component mass balances). If the set of columns is unknown then the mathematical formulation results in a non-convex Mixed-Integer Nonlinear Programming MINLP model [4].

The nonconvexities of the formulation result from the form of the objective function (bilinear terms of flowrates and compositions) and the form of constraints (equalities that involve bilinear expressions of flowrates and compositions). These nonconvexities may result in a large number of local optimum solutions that can differ significantly from each other as indicated in [4], which makes the search for the global optimum of major importance.

5.2 Test Problem 1 : Propane, Isobutane, n-Butane Nonsharp Separation

This test problem, which is a modified version of example 1 of [4], involves a three component feed mixture that has to be separated into two three component products. To avoid the distribution of nonkey components the recoveries of the key components were set to be greater than 0.85. The problem is formulated as a nonlinear programming problem NLP.

5.2.1 Problem Formulation

$$
MIN \quad a_{0,1} + \{a_{1,1} + a_{2,1}r_{A,1}^{lk} + a_{3,1}r_{B,1}^{hk} + b_{A,1}x_{A,5} + b_{B,1}x_{B,5}\}F_5
$$
$$
+ \ a_{0,2} + \{a_{1,2} + a_{2,2}r_{B,2}^{lk} + a_{3,2}r_{C,2}^{hk} + b_{A,2}x_{A,13} + b_{B,2}x_{B,13}\}F_{13}
$$

Subject to

$$
F_1 + F_2 + F_3 + F_4 = 300
$$

$$F_6 - F_7 - F_8 = 0$$

$$F_9 - F_{10} - F_{11} - F_{12} = 0$$

$$F_{14} - F_{15} - F_{16} - F_{17} = 0$$

$$F_{18} - F_{19} - F_{20} = 0$$

$$F_6 x_{A,6} - r_{A,1}^{lk} f_{A,5} = 0$$

$$F_{14} x_{B,14} - r_{B,2}^{lk} f_{B,13} = 0$$

$$F_9 x_{B,9} - r_{B,1}^{hk} f_{B,5} = 0$$

$$F_{18} x_{C,18} - r_{C,2}^{hk} f_{C,13} = 0$$

$$f_{A,5} - F_5 x_{A,5} = 0$$

$$f_{B,5} - F_5 x_{B,5} = 0$$

$$f_{C,5} - F_5 x_{C,5} = 0$$

$$f_{A,13} - F_{13} x_{A,13} = 0$$

$$f_{B,13} - F_{13} x_{B,13} = 0$$

$$f_{C,13} - F_{13} x_{C,13} = 0$$

$$f_{A,5} - F_6 x_{A,6} - F_9 x_{A,9} = 0$$

$$f_{B,5} - F_6 x_{B,6} - F_9 x_{B,9} = 0$$

$$f_{C,5} - F_6 x_{C,6} - F_9 x_{C,9} = 0$$

$$f_{A,13} - F_{14} x_{A,14} - F_{18} x_{A,18} = 0$$

$$f_{B,13} - F_{14} x_{B,14} - F_{18} x_{B,18} = 0$$

$$f_{C,13} - F_{14} x_{C,14} - F_{18} x_{C,18} = 0$$

$$0.333 F_1 + F_{15} x_{A,14} - f_{A,5} = 0$$

$$0.333 F_1 + F_{15} x_{B,14} - f_{B,5} = 0$$

$$0.333 F_1 + F_{15} x_{C,14} - f_{C,5} = 0$$

$$0.333 F_2 + F_{10} x_{A,9} - f_{A,13} = 0$$

$$0.333 F_2 + F_{10} x_{B,9} - f_{B,13} = 0$$

$$0.333 F_2 + F_{10} x_{C,9} - f_{C,13} = 0$$

$$x_{C,6} = 0$$

$$x_{A,18} = 0$$

$$.333F_3 + F_7 x_{A,6} + F_{11} x_{A,9} + F_{16} x_{A,14} + F_{19} x_{A,18} = 30$$

$$.333F_3 + F_7 x_{B,6} + F_{11} x_{B,9} + F_{16} x_{B,14} + F_{19} x_{B,18} = 50$$

$$.333F_3 + F_7 x_{C,6} + F_{11} x_{C,9} + F_{16} x_{C,14} + F_{19} x_{C,18} = 30$$

$$x_{A,5} + x_{B,5} + x_{C,5} = 1$$

$$x_{A,6} + x_{B,6} + x_{C,6} = 1$$

$$x_{A,9} + x_{B,9} + x_{C,9} = 1$$

$$x_{A,13} + x_{B,13} + x_{C,13} = 1$$

$$x_{A,14} + x_{B,14} + x_{C,14} = 1$$

$$x_{A,18} + x_{B,18} + x_{C,18} = 1$$

$$0.85 \le r_{i,j}^{lk}, r_{i,j}^{hk} \le 1$$

5.2.2 Data

Coefficient	Column I	Column II
$a_{0,i}$	0.23947	0.75835
$a_{1,i}$	-0.0139904	-0.0661588
$a_{2,i}$	0.0093514	0.0338147
$a_{3,i}$	0.0077308	0.0373349
$b_{A,i}$	-0.0005719	0.0016371
$b_{B,i}$	0.0042656	0.0288996

5.2.3 Problem Statistics

No. of Continuous Variables	48
No. of Binary Variables	–
No. of Linear Constraints	13
No. of Nonlinear Constraints	25

5.2.4 Best Known Solution

The best known solution for this Test Problem has an objective function value of 1.5671. The values of the nonzero variables are shown below :

$$F_2 = F_{13} \; = \; 85.714$$
$$F_3 \; = \; 77.143$$
$$F_4 \; = \; 137.143$$
$$F_5 = F_{14} = F_{15} \; = \; 57.143$$
$$F_6 = F_8 \; = \; 24.286$$
$$F_9 = F_{11} \; = \; 32.857$$
$$F_{18} = F_{20} \; = \; 28.571$$
$$f_{A,5} \; = \; 28.571$$
$$f_{B,5} \; = \; 24.286$$
$$f_{C,5} \; = \; 4.286$$
$$f_{A,13} = f_{B,13} = f_{C,13} \; = \; 28.571$$
$$x_{A,5} \; = \; 0.5$$
$$x_{B,5} \; = \; 0.425$$
$$x_{C,5} \; = \; 0.075$$
$$x_{A,6} \; = \; 1.0$$
$$x_{A,9} = x_{C,9} \; = \; 0.13$$
$$x_{B,9} \; = \; 0.739$$
$$x_{A,13} = x_{A,13} = x_{A,13} \; = \; 0.333$$
$$x_{A,14} \; = \; 0.5$$
$$x_{A,14} \; = \; 0.425$$
$$x_{A,14} \; = \; 0.075$$
$$x_{A,18} \; = \; 0.15$$
$$x_{A,18} \; = \; 0.85$$
$$r^{lk}_{A,1} = r^{lk}_{B,2} = r^{hk}_{C,2} \; = \; 0.85$$
$$r^{hk}_{B,1} \; = \; 1.0$$

5.3 Test Problem 2 : Propane, Isobutane, n-Butane, Isopentane Separation

In this test problem, a four component mixture has to be separated into two multicomponent products (see example 3 of [4]). The resulting mathematical formulation is a nonconvex NLP. The number of local optimal solutions increases rapidly with the number of columns (in this example three columns exist) and these solutions differ widely in structure and cost (variations of up to 25% in the cost of nonsharp sequences).

5.3.1 Problem Formulation

$$MIN \quad a_{0,1} + \{a_{1,1} + a_{2,1}r_{A,1}^{lk} + a_{3,1}r_{B,1}^{hk} + b_{A,1}x_{A,6} + b_{B,1}x_{B,6} + b_{C,1}x_{C,6}\}F_6$$

$$+ \ a_{0,2} + \{a_{1,2} + a_{2,2}r_{B,2}^{lk} + a_{3,2}r_{C,2}^{hk} + b_{A,2}x_{A,15} + b_{B,2}x_{B,15} + b_{C,1}x_{C,15}\}F_{15}$$

$$+ \ a_{0,3} + \{a_{1,3} + a_{2,3}r_{C,3}^{lk} + a_{3,3}r_{D,3}^{hk} + b_{A,3}x_{A,24} + b_{B,3}x_{B,24} + b_{C,3}x_{C,24}\}F_{24}$$

Subject to

$$F_1 + F_2 + F_3 + F_4 + F_5 = 600$$

$$F_7 - F_8 - F_9 = 0$$

$$F_{10} - F_{11} - F_{12} - F_{13} - F_{14} = 0$$

$$F_{16} - F_{17} - F_{18} - F_{19} = 0$$

$$F_{20} - F_{21} - F_{22} - F_{23} = 0$$

$$F_{25} - F_{26} - F_{27} - F_{28} - F_{29} = 0$$

$$F_{30} - F_{31} - F_{32} = 0$$

$$F_7 x_{A,7} - r_{A,1}^{lk} f_{A,6} = 0$$

$$F_{16} x_{B,16} - r_{B,2}^{lk} f_{B,15} = 0$$

$$F_{25} x_{C,25} - r_{C,3}^{lk} f_{C,24} = 0$$

$$F_{10} x_{B,10} - r_{B,1}^{hk} f_{B,6} = 0$$

$$F_{20}x_{C,20} - r_{C,2}^{hk}f_{C,15} = 0$$

$$F_{30}x_{D,30} - r_{D,3}^{hk}f_{D,24} = 0$$

$$f_{A,6} - F_6 x_{A,6} = 0$$

$$f_{B,6} - F_6 x_{B,6} = 0$$

$$f_{C,6} - F_6 x_{C,6} = 0$$

$$f_{D,6} - F_6 x_{D,6} = 0$$

$$f_{A,15} - F_{15} x_{A,15} = 0$$

$$f_{B,15} - F_{15} x_{B,15} = 0$$

$$f_{C,15} - F_{15} x_{C,15} = 0$$

$$f_{D,15} - F_{15} x_{D,15} = 0$$

$$f_{A,24} - F_{24} x_{A,24} = 0$$

$$f_{B,24} - F_{24} x_{B,24} = 0$$

$$f_{C,24} - F_{24} x_{C,24} = 0$$

$$f_{D,24} - F_{24} x_{D,24} = 0$$

$$f_{A,6} - F_7 x_{A,7} - F_{10} x_{A,10} = 0$$

$$f_{B,6} - F_7 x_{B,7} - F_{10} x_{B,10} = 0$$

$$f_{C,6} - F_7 x_{C,7} - F_{10} x_{C,10} = 0$$

$$f_{D,6} - F_7 x_{D,7} - F_{10} x_{D,10} = 0$$

$$f_{A,15} - F_{16} x_{A,16} - F_{20} x_{A,20} = 0$$

$$f_{B,15} - F_{16} x_{B,16} - F_{20} x_{B,20} = 0$$

$$f_{C,15} - F_{16} x_{C,16} - F_{20} x_{C,20} = 0$$

$$f_{D,15} - F_{16} x_{D,16} - F_{20} x_{D,20} = 0$$

$$f_{A,24} - F_{25} x_{A,25} - F_{30} x_{A,30} = 0$$

$$f_{B,24} - F_{25} x_{B,25} - F_{30} x_{B,30} = 0$$

$$f_{C,24} - F_{25} x_{C,25} - F_{30} x_{C,30} = 0$$

$$f_{D,24} - F_{25} x_{D,25} - F_{30} x_{D,30} = 0$$

$$0.250F_1 + F_{17} x_{A,17} + F_{26} x_{A,26} - f_{A,6} = 0$$

$$0.333F_1 + F_{17}x_{B,17} + F_{26}x_{B,26} - f_{B,6} = 0$$

$$0.167F_1 + F_{17}x_{C,17} + F_{26}x_{C,26} - f_{C,6} = 0$$

$$0.250F_1 + F_{17}x_{D,17} + F_{26}x_{D,26} - f_{A,6} = 0$$

$$0.250F_2 + F_{11}x_{A,11} + F_{27}x_{A,27} - f_{A,15} = 0$$

$$0.333F_2 + F_{11}x_{B,11} + F_{27}x_{B,27} - f_{B,15} = 0$$

$$0.167F_2 + F_{11}x_{C,11} + F_{27}x_{C,27} - f_{C,15} = 0$$

$$0.250F_2 + F_{11}x_{D,11} + F_{27}x_{D,27} - f_{A,15} = 0$$

$$0.250F_3 + F_{12}x_{A,12} + F_{21}x_{A,21} - f_{A,24} = 0$$

$$0.333F_3 + F_{12}x_{B,12} + F_{21}x_{B,21} - f_{B,24} = 0$$

$$0.167F_3 + F_{12}x_{C,12} + F_{21}x_{C,21} - f_{C,24} = 0$$

$$0.250F_3 + F_{12}x_{D,12} + F_{21}x_{D,21} - f_{A,24} = 0$$

$$x_{C,7} = 0$$

$$x_{D,7} = 0$$

$$x_{D,16} = 0$$

$$x_{A,20} = 0$$

$$x_{A,30} = 0$$

$$x_{B,30} = 0$$

$$.250F_4 + F_8x_{A,7} + F_{13}x_{A,10} + F_{18}x_{A,16} + F_{22}x_{A,20} + F_{28}x_{A,25} + F_{31}x_{A,30} = 50$$

$$.222F_4 + F_8x_{B,7} + F_{13}x_{B,10} + F_{18}x_{B,16} + F_{22}x_{B,20} + F_{28}x_{B,25} + F_{31}x_{B,30} = 100$$

$$.167F_4 + F_8x_{C,7} + F_{13}x_{C,10} + F_{18}x_{C,16} + F_{22}x_{C,20} + F_{28}x_{C,25} + F'_{31}x_{C,30} = 40$$

$$.250F_4 + F_8x_{D,7} + F_{13}x_{D,10} + F_{18}x_{D,16} + F_{22}x_{D,20} + F_{28}x_{D,25} + F_{31}x_{D,30} = 100$$

$$x_{A,6} + x_{B,6} + x_{C,6} + x_{D,6} = 1$$

$$x_{A,7} + x_{B,7} + x_{C,7} + x_{D,7} = 1$$

$$x_{A,10} + x_{B,10} + x_{C,10} + x_{D,10} = 1$$

$$x_{A,15} + x_{B,15} + x_{C,15} + x_{D,15} = 1$$

$$x_{A,16} + x_{B,16} + x_{C,16} + x_{D,16} = 1$$

$$x_{A,20} + x_{B,20} + x_{C,20} + x_{D,20} = 1$$

$$x_{A,24} + x_{B,24} + x_{C,24} + x_{D,24} = 1$$

$$x_{A,25} + x_{B,25} + x_{C,25} + x_{D,25} = 1$$

$$x_{A,30} + x_{B,30} + x_{C,30} + x_{D,30} = 1$$

$$0.85 \leq r_{i,j}^{lk}, r_{i,j}^{hk} \leq 1$$

5.3.2 Data

Coefficient	Column I	Column II	Column III
$a_{0,i}$	0.31569	0.96926	0.40281
$a_{1,i}$	-0.0112812	-0.0413393	-0.0119785
$a_{2,i}$	0.0072698	0.0228203	0.0082055
$a_{3,i}$	0.0064241	0.0257035	0.009819
$b_{A,i}$	0.0016446	0.0015625	-0.001748
$b_{B,i}$	0.0018611	0.0091604	-0.0002583
$b_{C,i}$	0.001262	0.0076758	-0.0004691

5.3.3 Problem Statistics

No. of Continuous Variables	86
No. of Binary Variables	–
No. of Linear Constraints	22
No. of Nonlinear Constraints	46

5.3.4 Best Known Solution

The best known solution for this Test Problem has an objective function value of 2.9852. The values of the nonzero variables are shown below :

$$F_1 = 47.059$$

$$F_2 = F_{15} = 85.714$$

$$F_3 = 117.647$$

$$F_4 = 180.084$$

$$F_5 = 169.496$$

$$F_6 = 94.916$$

$$F_7 = F_9 = 28.214$$

$$F_{10} = F_{13} = 66.702$$

$$F_{16} = F_{17} = 47.857$$

$$F_{20} = F_{21} = 37.857$$

$$F_{24} = 155.504$$

$$F_{25} = F_{29} = 112.290$$

$$F_{30} = F_{31} = 43.214$$

$$f_{A,6} = 33.193$$

$$f_{B,6} = 39.972$$

$$f_{C,6} = 9.986$$

$$f_{D,6} = 11.765$$

$$f_{A,15} = f_{D,15} = 21.479$$

$$f_{B,15} = 28.571$$

$$f_{C,15} = 14.286$$

$$f_{A,24} = 29.412$$

$$f_{B,24} = 43.501$$

$$f_{C,24} = 31.751$$

$$f_{D,24} = 50.840$$

$$x_{A,6} = 0.35$$

$$x_{B,6} = 0.421$$

$$x_{C,6} = 0.105$$

$$x_{D,6} = 0.124$$

$$x_{A,7} = 1.0$$

$$x_{A,10} = 0.075$$

$$x_{B,10} = 0.599$$

$$x_{C,10} = 0.15$$

$$x_{D,10} = 0.176$$

$$x_{A,15} = x_{D,15} = 0.25$$

$$x_{B,15} = 0.333$$

$$x_{C,15} = 0.167$$

$$x_{A,16} = 0.448$$

$$x_{B,16} = 0.507$$

$$x_{C,16} = 0.045$$

$$x_{B,20} = 0.113$$

$$x_{C,20} = 0.321$$

$$x_{D,20} = 0.566$$

$$x_{A,24} = 0.189$$

$$x_{B,24} = 0.28$$

$$x_{C,24} = 0.204$$

$$x_{D,24} = 0.327$$

$$x_{A,25} = 0.262$$

$$x_{B,25} = 0.387$$

$$x_{C,25} = 0.283$$

$$x_{D,25} = 0.068$$

$$x_{D,30} = 1.0$$

$$r_{A,1}^{lk} = r_{B,2}^{lk} = r_{C,2}^{hk} = r_{D,3}^{hk} = 0.85$$

$$r_{B,1}^{hk} = r_{C,3}^{lk} = 1.0$$

5.4 Test Problem 3 : Blending/Pooling/Separation Problem

This test problem is taken from [73] and involves a three-component feed mixture that is to be separated into two multicomponent products by using separators and splitting/blending/pooling. The cost of each separator depends linearly on the flowrate through the separator, and the constraints correspond to mass balances around the various splitters, separators and mixers.

5.4.1 Problem Formulation

$$MIN \qquad 0.9979 + 0.00432F_5 + 0.01517F_{13}$$

Subject to

$$F_1 + F_2 + F_3 + F_4 = 300$$

$$F_6 - F_7 - F_8 = 0$$

$$F_9 - F_{10} - F_{11} - F_{12} = 0$$

$$F_{14} - F_{15} - F_{16} - F_{17} = 0$$

$$F_{18} - F_{19} - F_{20} = 0$$

$$F_5 x_{A,5} - F_6 x_{A,6} - F_9 x_{A,9} = 0$$

$$F_5 x_{B,5} - F_6 x_{B,6} - F_9 x_{B,9} = 0$$

$$F_5 x_{C,5} - F_6 x_{C,6} - F_9 x_{C,9} = 0$$

$$F_{13} x_{A,13} - F_{14} x_{A,14} - F_{18} x_{A,18} = 0$$

$$F_{13} x_{B,13} - F_{14} x_{B,14} - F_{18} x_{B,18} = 0$$

$$F_{13} x_{C,13} - F_{14} x_{C,14} - F_{18} x_{C,18} = 0$$

$$0.333F_1 + F_{15} x_{A,14} - F_5 x_{A,5} = 0$$

$$0.333F_1 + F_{15} x_{B,14} - F_5 x_{B,5} = 0$$

$$0.333F_1 + F_{15} x_{C,14} - F_5 x_{C,5} = 0$$

$$0.333F_2 + F_{10} x_{A,9} - F_{13} x_{A,13} = 0$$

$$0.333F_2 + F_{10} x_{B,9} - F_{13} x_{B,13} = 0$$

$$0.333F_2 + F_{10}x_{C,9} - F_{13}x_{C,13} = 0$$

$$.333F_3 + F_7x_{A,6} + F_{11}x_{A,9} + F_{16}x_{A,14} + F_{19}x_{A,18} = 30$$

$$.333F_3 + F_7x_{B,6} + F_{11}x_{B,9} + F_{16}x_{B,14} + F_{19}x_{B,18} = 50$$

$$.333F_3 + F_7x_{C,6} + F_{11}x_{C,9} + F_{16}x_{C,14} + F_{19}x_{C,18} = 30$$

$$x_{A,5} + x_{B,5} + x_{C,5} = 1$$

$$x_{A,6} + x_{B,6} + x_{C,6} = 1$$

$$x_{A,9} + x_{B,9} + x_{C,9} = 1$$

$$x_{A,13} + x_{B,13} + x_{C,13} = 1$$

$$x_{A,14} + x_{B,14} + x_{C,14} = 1$$

$$x_{A,18} + x_{B,18} + x_{C,18} = 1$$

$$x_{B,6} = 0$$

$$x_{C,6} = 0$$

$$x_{A,9} = 0$$

$$x_{C,14} = 0$$

$$x_{A,18} = 0$$

$$x_{B,18} = 0$$

5.4.2   Problem Statistics

No. of Continuous Variables	38
No. of Binary Variables	–
No. of Linear Constraints	17
No. of Nonlinear Constraints	15

5.4.3   Best Known Solution

The best known solution for this Test Problem has an objective function value of 1.8639. The values of the nonzero variables are shown below :

$$F_1 = F_5 = 60$$

$$F_3 = 90$$

$$F_4 = 150$$

$$F_6 = F_8 = 20$$

$$F_9 = F_{10} = F_{13} = 40$$

$$F_{14} = F_{16} = F_{18} = F_{20} = 20$$

$$x_{A,5} = x_{B,5} = x_{C,5} = 0.333$$

$$x_{A,6} = x_{B,14} = x_{C,18} = 1.0$$

$$x_{B,9} = x_{C,9} = x_{B,13} = x_{C,13} = 0.5$$

5.5 Test Problem 4 : Three Component Separation - MINLP

In this test problem (example 1 of [4]), the same feed mixture as in test problem 1 has to be separated into a different set of products. The composition though, of the desired products is different. The mathematical formulation involves the existence or nonexistence of the columns as binary variables explicitly and corresponds to a nonconvex MINLP.

5.5.1 Problem Formulation

MIN

$$a_{0,1}y_1 + \{a_{1,1} + a_{2,1}r_{A,1}^{lk} + a_{3,1}r_{B,1}^{hk} + b_{A,1}x_{A,5} + b_{B,1}x_{B,5}\}F_5$$
$$+ \ a_{0,2}y_2 + \{a_{1,2} + a_{2,2}r_{B,13}^{lk} + a_{3,2}r_{C,13}^{hk} + b_{A,2}x_{A,13} + b_{B,2}x_{B,13}\}F_{13}$$

Subject to

$$F_1 + F_2 + F_3 + F_4 = 300$$

$$F_6 - F_7 - F_8 = 0$$

$$F_9 - F_{10} - F_{11} - F_{12} = 0$$

$$F_{14} - F_{15} - F_{16} - F_{17} = 0$$

$$F_{18} - F_{19} - F_{20} = 0$$

$$F_6 x_{A,6} - r_{A,1}^{lk} f_{A,5} = 0$$

$$F_{14}x_{B,14} - r_{B,2}^{lk}f_{B,13} = 0$$

$$F_9 x_{B,9} - r_{B,1}^{hk}f_{B,5} = 0$$

$$F_{18}x_{C,18} - r_{C,2}^{hk}f_{C,13} = 0$$

$$f_{A,5} - F_5 x_{A,5} = 0$$

$$f_{B,5} - F_5 x_{B,5} = 0$$

$$f_{C,5} - F_5 x_{C,5} = 0$$

$$f_{A,13} - F_{13}x_{A,13} = 0$$

$$f_{B,13} - F_{13}x_{B,13} = 0$$

$$f_{C,13} - F_{13}x_{C,13} = 0$$

$$f_{A,5} - F_6 x_{A,6} - F_9 x_{A,9} = 0$$

$$f_{B,5} - F_6 x_{B,6} - F_9 x_{B,9} = 0$$

$$f_{C,5} - F_6 x_{C,6} - F_9 x_{C,9} = 0$$

$$f_{A,13} - F_{14}x_{A,14} - F_{18}x_{A,18} = 0$$

$$f_{B,13} - F_{14}x_{B,14} - F_{18}x_{B,18} = 0$$

$$f_{C,13} - F_{14}x_{C,14} - F_{18}x_{C,18} = 0$$

$$0.333F_1 + F_{15}x_{A,14} - f_{A,5} = 0$$

$$0.333F_1 + F_{15}x_{B,14} - f_{B,5} = 0$$

$$0.333F_1 + F_{15}x_{C,14} - f_{C,5} = 0$$

$$0.333F_2 + F_{10}x_{A,9} - f_{A,13} = 0$$

$$0.333F_2 + F_{10}x_{B,9} - f_{B,13} = 0$$

$$0.333F_2 + F_{10}x_{C,9} - f_{C,13} = 0$$

$$x_{C,6} = 0$$

$$x_{A,18} = 0$$

$$.333F_3 + F_7 x_{A,6} + F_{11}x_{A,9} + F_{16}x_{A,14} + F_{19}x_{A,18} = 80$$

$$.333F_3 + F_7 x_{B,6} + F_{11}x_{B,9} + F_{16}x_{B,14} + F_{19}x_{B,18} = 30$$

$$.333F_3 + F_7 x_{C,6} + F_{11}x_{C,9} + F_{16}x_{C,14} + F_{19}x_{C,18} = 20$$

$$x_{A,5} + x_{B,5} + x_{C,5} = 1$$

$$x_{A,6} + x_{B,6} + x_{C,6} = 1$$

$$x_{A,9} + x_{B,9} + x_{C,9} = 1$$

$$x_{A,13} + x_{B,13} + x_{C,13} = 1$$

$$x_{A,14} + x_{B,14} + x_{C,14} = 1$$

$$x_{A,18} + x_{B,18} + x_{C,18} = 1$$

$$F_5 - 300y_1 \leq 0$$

$$F_{13} - 300y_2 \leq 0$$

$$0.85 \leq r_{i,j}^{lk}, r_{i,j}^{hk} \leq 1$$

5.5.2 Data

Coefficient	Column I	Column II
$a_{0,i}$	0.23947	0.75835
$a_{1,i}$	-0.0139904	-0.0661588
$a_{2,i}$	0.0093514	0.0338147
$a_{3,i}$	0.0077308	0.0373349
$b_{A,i}$	-0.0005719	0.0016371
$b_{B,i}$	0.0042656	0.0288996

5.5.3 Problem Statistics

No. of Continuous Variables	48
No. of Binary Variables	2
No. of Linear Constraints	13
No. of Nonlinear Constraints	25

5.5.4 Best Known Solution

The best known solution for this Test Problem has an objective function value of 0.626. This solution involves only column I and the values of the nonzero ariables are shown below :

$$y_1 = 1$$
$$F_1 = F_5 = 211.765$$
$$F_3 = 60$$
$$F_4 = 28.235$$
$$F_6 = F_7 = 70$$
$$F_9 = F_{12} = 141.765$$
$$f_{A,5} = f_{B,5} = f_{C,5} = 70.588$$
$$x_{A,5} = x_{A,5} = x_{A,5} = 0.333$$
$$x_{A,6} = 0.857$$
$$x_{B,6} = 0.143$$
$$x_{A,9} = 0.075$$
$$x_{B,9} = 0.427$$
$$x_{C,9} = 0.498$$
$$r_{A,1}^{lk} = 0.85$$
$$r_{B,1}^{hk} = 0.858$$

5.6   Test Problem 5 : Four Component Separation - MINLP

A four component feed stream is to be separated into two products (see example 4 of [4]). The problem is formulated as a mixed-integer nonlinear programme MINLP.

5.6.1   Problem Formulation

MIN
$$a_{0,1}y_1 + \{a_{1,1} + a_{2,1}r_{A,1}^{lk} + a_{3,1}r_{B,1}^{hk} + b_{A,1}x_{A,6} + b_{B,1}x_{B,6} + b_{C,1}x_{C,6}\}F_6$$
$$+ \ a_{0,2}y_2 + \{a_{1,2} + a_{2,2}r_{B,2}^{lk} + a_{3,2}r_{C,2}^{hk} + b_{A,2}x_{A,15} + b_{B,2}x_{B,15} + b_{C,1}x_{C,15}\}F_{15}$$
$$+ \ a_{0,3}y_3 + \{a_{1,3} + a_{2,3}r_{C,3}^{lk} + a_{3,3}r_{D,3}^{hk} + b_{A,3}x_{A,24} + b_{B,3}x_{B,24} + b_{C,3}x_{C,24}\}F_{24}$$

Subject to

$$F_1 + F_2 + F_3 + F_4 + F_5 = 600$$

$$F_7 - F_8 - F_9 = 0$$

$$F_{10} - F_{11} - F_{12} - F_{13} - F_{14} = 0$$

$$F_{16} - F_{17} - F_{18} - F_{19} = 0$$

$$F_{20} - F_{21} - F_{22} - F_{23} = 0$$

$$F_{25} - F_{26} - F_{27} - F_{28} - F_{29} = 0$$

$$F_{30} - F_{31} - F_{32} = 0$$

$$F_7 x_{A,7} - r_{A,1}^{lk} f_{A,6} = 0$$

$$F_{16} x_{B,16} - r_{B,2}^{lk} f_{B,15} = 0$$

$$F_{25} x_{C,25} - r_{C,3}^{lk} f_{C,24} = 0$$

$$F_{10} x_{B,10} - r_{B,1}^{hk} f_{B,6} = 0$$

$$F_{20} x_{C,20} - r_{C,2}^{hk} f_{C,15} = 0$$

$$F_{30} x_{D,30} - r_{D,3}^{hk} f_{D,24} = 0$$

$$f_{A,6} - F_6 x_{A,6} = 0$$

$$f_{B,6} - F_6 x_{B,6} = 0$$

$$f_{C,6} - F_6 x_{C,6} = 0$$

$$f_{D,6} - F_6 x_{D,6} = 0$$

$$f_{A,15} - F_{15} x_{A,15} = 0$$

$$f_{B,15} - F_{15} x_{B,15} = 0$$

$$f_{C,15} - F_{15} x_{C,15} = 0$$

$$f_{D,15} - F_{15} x_{D,15} = 0$$

$$f_{A,24} - F_{24} x_{A,24} = 0$$

$$f_{B,24} - F_{24} x_{B,24} = 0$$

$$f_{C,24} - F_{24} x_{C,24} = 0$$

$$f_{D,24} - F_{24} x_{D,24} = 0$$

$$f_{A,6} - F_7 x_{A,7} - F_{10} x_{A,10} = 0$$

$$f_{B,6} - F_7 x_{B,7} - F_{10} x_{B,10} = 0$$

$$f_{C,6} - F_7 x_{C,7} - F_{10} x_{C,10} = 0$$

$$f_{D,6} - F_7 x_{D,7} - F_{10} x_{D,10} = 0$$

$$f_{A,15} - F_{16} x_{A,16} - F_{20} x_{A,20} = 0$$

$$f_{B,15} - F_{16} x_{B,16} - F_{20} x_{B,20} = 0$$

$$f_{C,15} - F_{16} x_{C,16} - F_{20} x_{C,20} = 0$$

$$f_{D,15} - F_{16} x_{D,16} - F_{20} x_{D,20} = 0$$

$$f_{A,24} - F_{25} x_{A,25} - F_{30} x_{A,30} = 0$$

$$f_{B,24} - F_{25} x_{B,25} - F_{30} x_{B,30} = 0$$

$$f_{C,24} - F_{25} x_{C,25} - F_{30} x_{C,30} = 0$$

$$f_{D,24} - F_{25} x_{D,25} - F_{30} x_{D,30} = 0$$

$$0.250 F_1 + F_{17} x_{A,17} + F_{26} x_{A,26} - f_{A,6} = 0$$

$$0.333 F_1 + F_{17} x_{B,17} + F_{26} x_{B,26} - f_{B,6} = 0$$

$$0.167 F_1 + F_{17} x_{C,17} + F_{26} x_{C,26} - f_{C,6} = 0$$

$$0.250 F_1 + F_{17} x_{D,17} + F_{26} x_{D,26} - f_{A,6} = 0$$

$$0.250 F_2 + F_{11} x_{A,11} + F_{27} x_{A,27} - f_{A,15} = 0$$

$$0.333 F_2 + F_{11} x_{B,11} + F_{27} x_{B,27} - f_{B,15} = 0$$

$$0.167 F_2 + F_{11} x_{C,11} + F_{27} x_{C,27} - f_{C,15} = 0$$

$$0.250 F_2 + F_{11} x_{D,11} + F_{27} x_{D,27} - f_{A,15} = 0$$

$$0.250 F_3 + F_{12} x_{A,12} + F_{21} x_{A,21} - f_{A,24} = 0$$

$$0.333 F_3 + F_{12} x_{B,12} + F_{21} x_{B,21} - f_{B,24} = 0$$

$$0.167 F_3 + F_{12} x_{C,12} + F_{21} x_{C,21} - f_{C,24} = 0$$

$$0.250 F_3 + F_{12} x_{D,12} + F_{21} x_{D,21} - f_{A,24} = 0$$

$$x_{C,7} = 0$$

$$x_{D,7} = 0$$

$$x_{D,16} = 0$$

$$x_{A,20} = 0$$

$$x_{A,30} = 0$$

$$x_{B,30} = 0$$

$$.250F_4 + F_8 x_{A,7} + F_{13} x_{A,10} + F_{18} x_{A,16} + F_{22} x_{A,20} + F_{28} x_{A,25} + F_{31} x_{A,30} = 75$$

$$.222F_4 + F_8 x_{B,7} + F_{13} x_{B,10} + F_{18} x_{B,16} + F_{22} x_{B,20} + F_{28} x_{B,25} + F_{31} x_{B,30} = 100$$

$$.167F_4 + F_8 x_{C,7} + F_{13} x_{C,10} + F_{18} x_{C,16} + F_{22} x_{C,20} + F_{28} x_{C,25} + F_{31} x_{C,30} = 40$$

$$.250F_4 + F_8 x_{D,7} + F_{13} x_{D,10} + F_{18} x_{D,16} + F_{22} x_{D,20} + F_{28} x_{D,25} + F_{31} x_{D,30} = 100$$

$$x_{A,6} + x_{B,6} + x_{C,6} + x_{D,6} = 1$$

$$x_{A,7} + x_{B,7} + x_{C,7} + x_{D,7} = 1$$

$$x_{A,10} + x_{B,10} + x_{C,10} + x_{D,10} = 1$$

$$x_{A,15} + x_{B,15} + x_{C,15} + x_{D,15} = 1$$

$$x_{A,16} + x_{B,16} + x_{C,16} + x_{D,16} = 1$$

$$x_{A,20} + x_{B,20} + x_{C,20} + x_{D,20} = 1$$

$$x_{A,24} + x_{B,24} + x_{C,24} + x_{D,24} = 1$$

$$x_{A,25} + x_{B,25} + x_{C,25} + x_{D,25} = 1$$

$$x_{A,30} + x_{B,30} + x_{C,30} + x_{D,30} = 1$$

$$F_6 - 600y_1 \leq 0$$

$$F_{15} - 600y_2 \leq 0$$

$$F_{24} - 600y_3 \leq 0$$

$$0.85 \leq r_{i,j}^{lk}, r_{i,j}^{hk} \leq 1$$

5.6.2 Data

5.6.3 Problem Statistics

5.6.4 Best Known Solution

The best known solution for this Test Problem has an objective function value of 2.579 and has only two columns (II and III). The values of the nonzero variables

Coefficient	Column I	Column II	Column III
$a_{0,i}$	0.31569	0.96926	0.40281
$a_{1,i}$	-0.0112812	-0.0413393	-0.0119785
$a_{2,i}$	0.0072698	0.0228203	0.0082055
$a_{3,i}$	0.0064241	0.0257035	0.009819
$b_{A,i}$	0.0016446	0.0015625	-0.001748
$b_{B,i}$	0.0018611	0.0091604	-0.0002583
$b_{C,i}$	0.001262	0.0076758	-0.0004691

No. of Continuous Variables	86
No. of Binary Variables	3
No. of Linear Constraints	22
No. of Nonlinear Constraints	46

are shown below :

$$y_2 = y_3 = 1$$
$$F_2 = F_{15} = 70.588$$
$$F_3 = 130.104$$
$$F_4 = 229.412$$
$$F_5 = 169.896$$
$$F_{16} = F_{18} = 42.941$$
$$F_{20} = F_{21} = 27.647$$
$$F_{24} = 157.751$$
$$F_{25} = F_{29} = 115.104$$
$$F_{30} = F_{31} = 42.647$$
$$f_{A,15} = f_{D,15} = 17.647$$
$$f_{B,15} = 23.529$$
$$f_{C,15} = 11.765$$
$$f_{A,24} = 32.526$$
$$f_{B,24} = 43.368$$

$$f_{C,24} = 31.684$$

$$f_{D,24} = 50.173$$

$$x_{A,15} = x_{D,15} = 0.25$$

$$x_{B,15} = 0.333$$

$$x_{C,15} = 0.167$$

$$x_{A,16} = 0.411$$

$$x_{B,16} = 0.548$$

$$x_{C,16} = 0.041$$

$$x_{C,20} = 0.362$$

$$x_{D,20} = 0.638$$

$$x_{A,24} = 0.206$$

$$x_{B,24} = 0.275$$

$$x_{C,24} = 0.201$$

$$x_{D,24} = 0.318$$

$$x_{A,25} = 0.283$$

$$x_{B,25} = 0.377$$

$$x_{C,25} = 0.275$$

$$x_{D,25} = 0.065$$

$$x_{D,30} = 1.0$$

$$r_{B,2}^{lk} = r_{C,3}^{lk} = 1.0$$

$$r_{C,2}^{hk} = r_{D,3}^{hk} = 0.85$$

Chapter 6

Pooling/Blending test problems

6.1 Problem Statement

In refinery and petrochemical processing problems it is generally necessary to model not only product flows but also properties of the streams as well. When streams are combined in a tank or pool, nonlinear relationships are often introduced if the pool is to be used in downstream blending or processing. In a number of blending problems, the qualities of the component streams contribute to the qualities of the blended product in a nonlinear and nonconvex manner.

Successive Linear Programming (SLP) techniques have been widely used in the industry for over 25 years. SLP algorithms solve nonlinear optimization problems via a sequence of linear programs. In [89] the first such algorithm, the Method of Approximation Programming (MAP), was presented. Recent work on the pooling problem includes [131], [158] , [14], and [73] who proposed a decomposition scheme that induces convex subproblems based on [74] and [3]. In his studies of the behaviour of linear programming LP models Haverly [99] defined a simple **pooling problem** that exhibits a number of local solutions that depend on the starting point.

6.2 Test Problem 1 : Haverly's Pooling Problem - Case I

6.2.1 Problem Formulation

$$\max \quad Profit = 9x + 15y - 6A - 16B - 10(Cx + Cy)$$

subject to

$$Px + Py - A - B = 0$$
$$x - Px - Cx = 0$$
$$y - Py - Cy = 0$$
$$p.Px + 2Cx - 2.5x \leq 0$$
$$p.Py + 2Cy - 1.5y \leq 0$$
$$p.Px + p.Py - 3A - B = 0$$
$$x \leq 100$$
$$y \leq 200$$

6.2.2 Problem Statistics

No. of Continuous Variables	9
No. of Linear Constraints	3
No. of Nonlinear Equality Constraints	1
No. of Nonlinear Inequality Constraints	2

6.2.3 Global Solution

The best known solution for this Test Problem has an objective function value of 400 and the values of the nonzero variables for this solution are :

$$p = 1$$
$$B = Py = Cy = 100$$
$$y = 200$$

6.3 Test Problem 2 : Haverly's Pooling Problem - Case II

6.3.1 Problem Formulation

The formulation for this problem is the same as for Test Problem 1 except that the upper bound on the product x is changed. The complete formulation is as follows :

$$\max \quad Profit = 9x + 15y - 6A - 16B - 10(Cx + Cy)$$

subject to

$$Px + Py - A - B = 0$$
$$x - Px - Cx = 0$$
$$y - Py - Cy = 0$$
$$p.Px + 2Cx - 2.5x \leq 0$$
$$p.Py + 2Cy - 1.5y \leq 0$$
$$p.Px + p.Py - 3A - B = 0$$
$$x \leq 600$$
$$y \leq 200$$

The problem statistics are the same as for Test Problem 1.

6.3.2 Global Solution

The best known solution for this Test Problem has an objective function value of 600 and the values of the nonzero variables for this solution are :

$$p = 3$$
$$A = Px = Cx = 300$$
$$x = 600$$

6.4 Test Problem 3 : Haverly's Pooling Problem - Case III

6.4.1 Problem Formulation

The formulation for this problem is the same as for Test Problem 1 except that in the objective function the cost of B is changed from 16 to 13. The new formulation is :

$$\max \quad Profit = 9x + 15y - 6A - 13B - 10(Cx + Cy)$$

subject to

$$Px + Py - A - B = 0$$
$$x - Px - Cx = 0$$
$$y - Py - Cy = 0$$
$$p.Px + 2Cx - 2.5x \leq 0$$
$$p.Py + 2Cy - 1.5y \leq 0$$
$$p.Px + p.Py - 3A - B = 0$$
$$x \leq 100$$
$$y \leq 200$$

The problem statistics are again the same as for the previous two problems.

6.4.2 Global Solution

The best known solution for this Test Problem has an objective function value of 750 and the values of the nonzero variables for this solution are :

$$p = 1.5$$
$$A = 50$$
$$B = 150$$
$$Py = 200$$
$$y = 200$$

Chapter 7

Heat Exchanger Network Synthesis test problems

7.1 Problem Statement

Heat exchanger network synthesis has been the subject of an intense research effort in the last two decades. Mathematical programming approaches have been developed and these procedures have been incorporated into several approaches to the heat exchanger network synthesis problem. Two classical subproblems of the heat exchanger synthesis problem are (a) the network configuration optimization problem and (b) the simultaneous matches-network optimization problem.

(i) Network configuration optimization problem

The optimization problem for the heat exchanger network configuration is stated as follows :

Given are: (a) a set of hot process streams and hot utilities $i \in H$; (b) a set of cold process streams and cold utilities, $j \in C$; (c) the inlet and outlet temperatures, the heat capacity flowrates, and individual heat transfer coefficients of each stream; (d) a minimum temperature approach $\Delta \mathbf{T_{min}}$; (e) the minimum utility consumption and the location of pinch points; and (f) a set of matches $(ij) \in MA$ satisfying the minimum number of units criterion, their heat loads $\mathbf{Q_{ij}}$ and heat transfer coefficients $\mathbf{U_{ij}}$. The objective is to determine the process stream matches and the heat exchanger network configuration that provides the globally minimum investment cost.

The network optimization problem can be formulated as a nonlinear mathematical programming (NLP) problem ([77]). Uncertainty arises because this nonlinear problem is nonconvex, and thus may have several local optima. When conventional solution techniques are used to solve this problem, the final solution depends upon the starting point.

(ii) Simultaneous matches-network optimization problem

The problem of determining simultaneously the best combination of process stream matches and network configuration can be stated as follows:

Given are: (a) a set of hot process streams and hot utilities $i \in H$, their inlet and outlet temperatures $\mathbf{T^i}$, $\mathbf{T^{O,i}}$, and heat capacity flowrates $\mathbf{F^i}$; b) a set of cold process streams and cold utilities $j \in C$, their inlet and outlet temperatures $\mathbf{T^j}$, $\mathbf{T^{O,j}}$, and heat capacity flowrates $\mathbf{F^j}$; (c) a minimum temperature approach $\mathbf{\Delta T_{min}}$; d) the overall heat transfer coefficients $\mathbf{U_{ij}}$ of each potential match (ij); (e) the minimum utility consumption and the location of pinch points; and (f) a maximum number of units $\mathbf{N_{max}}$.

The objective of this problem is to obtain the set of process stream matches and their heat loads that provide the minimum cost heat exchanger network configuration.

The simultaneous matches-network optimization problem is formulated as a mixed integer nonlinear programming (MINLP) problem that selects the set of matches, heat loads, and network configuration that feature the minimum investment cost from among all possible process stream matches and network configurations. The mathematical formulation is discussed in [76].

7.2 Test Problem 1 : Two-Unit Heat Exchanger Network - NLP

This problem involves determining the optimal heat exchanger network configuration for a system of two hot streams and one cold stream. The problem is taken from [76]. The objective of the problem is to obtain the optimal configuration of the cold stream so as to minimize the heat exchanger area investment cost. The formulation is based upon the heat exchanger network superstructure of [77].

7.2.1 Problem Formulation

$$\min \quad 1300 \left[\frac{1000}{0.05\left[\frac{2}{3}(\Delta T_{11}\Delta T_{12})^{1/2}+\frac{1}{6}(\Delta T_{11}+\Delta T_{12})\right]} \right]^{0.6} +$$
$$1300 \left[\frac{600}{0.05\left[\frac{2}{3}(\Delta T_{21}\Delta T_{22})^{1/2}+\frac{1}{6}(\Delta T_{21}+\Delta T_{22})\right]} \right]^{0.6}$$

subject to

$$
\begin{aligned}
f_1^I + f_2^I &= 10 \\
f_1^I + f_{12}^B - f_1^E &= 0 \\
f_2^I + f_{21}^B - f_2^E &= 0 \\
f_1^O + f_{21}^B - f_1^E &= 0 \\
f_2^O + f_{12}^B - f_2^E &= 0 \\
150 f_1^I + t_2^O f_{12}^B - t_1^I f_1^E &= 0 \\
150 f_2^I + t_1^O f_{21}^B - t_2^I f_2^E &= 0 \\
f_1^E(t_1^O - t_1^I) &= 1000 \\
f_2^E(t_2^O - t_2^I) &= 600 \\
\Delta T_{11} &= 500 - t_1^O \\
\Delta T_{12} &= 250 - t_1^I \\
\Delta T_{21} &= 350 - t_2^O \\
\Delta T_{22} &= 200 - t_2^I \\
\Delta T_{11},\ \Delta T_{12},\ \Delta T_{21},\ \Delta T_{22} &\geq 10 \\
0 \leq f_1^I,\ f_2^I,\ f_{12}^B,\ f_{21}^B,\ f_1^O,\ f_2^O &\leq 10 \\
2.941 \leq f_1^E &\leq 10 \\
3.158 \leq f_2^E &\leq 10 \\
150 \leq t_1^I &\leq 240
\end{aligned}
$$

$$250 \leq t_1^O \leq 490$$
$$150 \leq t_2^I \leq 190$$
$$210 \leq t_2^O \leq 340$$

7.2.2 Data

The stream data is given in Table 7.1, and information about the matches is given in Table 7.2. There are two matches within the heat exchanger network.

Table 7.1: Stream Data for Test Problem 1

STREAM	T in(K)	T out(K)	$FC_p(\frac{kW}{K})$
H1	500	250	4
H2	350	200	4
C1	150	310	10

$$\Delta T_{min} = 10^\circ K$$

Table 7.2: Match Data for Test Problem 1

MATCH	Q(kW)	U $(\frac{kW}{m^2K})$	A (m^2)
H1 C1	1000	0.05	207.357
H2 C1	600	0.05	137.23

Cost of Heat Exchangers $= \$1300 A^{0.6}$

7.2.3 Problem Statistics

This problem involves a system of 16 variables and 13 equality constraints. Bounds for each of the variables are also provided. Four of the constraints contain bilinear terms, and the remaining nine constraints contain only linear terms. The objective is a nonlinear but convex function.

7.2.4 Best Known Solution

The best known solution to this problem was reported in [76] . This solution involves a series arrangement of the exchangers, and features a total investment cost \$56,825. The best known solution is shown in Figure 7.1, and the values of the variables in the optimization problem are given in Tables 7.3 and 7.4.

Table 7.3: Flowrate and Temperature Levels at the Best Known Solution

$f_1^I = 0$	$f_2^I = 10$	$f_1^E = 10$	$f_2^E = 10$
$f_{12}^B = 10$	$f_{21}^B = 0$	$f_1^O = 10$	$f_2^O = 0$
$t_1^I = 210$	$t_2^I = 150$	$t_1^O = 310$	$t_2^O = 210$

Table 7.4: Temperature Differences at the Best Known Solution

$\Delta T_{11} = 190$	$\Delta T_{12} = 40$
$\Delta T_{21} = 140$	$\Delta T_{22} = 50$

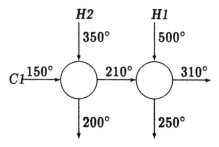

Figure 7.1: Best Known Solution to Test Problem 1

7.3 Test Problem 2 : Three-Unit Heat Exchanger Network - NLP

This example is taken from [76]. It involves obtaining an optimal heat exchanger network for a system of three hot streams and one cold stream. A nonlinear optimization problem has been formulated by creating a superstructure [77] containing all possible network configurations embedded within it. The nonlinear programming problem has as an objective the investment cost of the network. The goal of the optimization problem is to select from the superstructure the network structure with minimum investment cost.

7.3.1 Problem Formulation

$$minimize \quad 1300[\frac{Q_1}{U_1(\frac{2}{3}(\Delta T1_1 \Delta T2_1)^{1/2} + \frac{1}{6}(\Delta T1_1 + \Delta T2_1)}]^{0.6} +$$

$$1300[\frac{Q_2}{U_2(\frac{2}{3}(\Delta T1_2 \Delta T2_2)^{1/2} + \frac{1}{6}(\Delta T1_2 + \Delta T2_2)}]^{0.6} +$$

$$1300[\frac{Q_3}{U_3(\frac{2}{3}(\Delta T1_3 \Delta T2_3)^{1/2} + \frac{1}{6}(\Delta T1_3 + \Delta T2_3)}]^{0.6}$$

$$subject\ to$$

$$f_1^I + f_2^I + f_3^I = 45$$
$$f_1^I + f_{12}^B + f_{13}^B - f_1^E = 0$$
$$f_2^I + f_{21}^B + f_{23}^B - f_2^E = 0$$
$$f_3^I + f_{31}^B + f_{32}^B - f_3^E = 0$$
$$f_1^O + f_{21}^B + f_{31}^B - f_1^E = 0$$
$$f_2^O + f_{12}^B + f_{32}^B - f_2^E = 0$$
$$f_3^O + f_{13}^B + f_{23}^B - f_3^E = 0$$
$$100 f_1^I + t_2^O f_{12}^B + t_3^O f_{13}^B - t_1^I f_1^E = 0$$
$$100 f_2^I + t_1^O f_{21}^B + t_3^O f_{23}^B - t_2^I f_2^E = 0$$
$$100 f_3^I + t_1^O f_{31}^B + t_2^O f_{32}^B - t_4^I f_3^E = 0$$

$$f_1^E(t_1^O - t_1^I) = Q_1$$

$$f_2^E(t_2^O - t_2^I) = Q_2$$

$$f_3^E(t_3^O - t_3^I) = Q_3$$

$$\Delta T1_1 = T_1^I - t_1^O$$

$$\Delta T2_1 = T_1^O - t_1^I$$

$$\Delta T1_2 = T_2^I - t_2^O$$

$$\Delta T2_2 = T_2^O - t_2^I$$

$$\Delta T1_3 = T_3^I - t_3^O$$

$$\Delta T2_3 = T_3^O - t_3^I$$

$$\Delta T1_1, \Delta T2_1 \geq \Delta T_{min}$$

$$\Delta T1_2, \Delta T2_2 \geq \Delta T_{min}$$

$$\Delta T1_3, \Delta T2_3 \geq \Delta T_{min}$$

$$0 \leq f_1^I \leq 45$$

$$0 \leq f_2^I \leq 45$$

$$0 \leq f_3^I \leq 45$$

$$0 \leq f_{12}^B \leq 45$$

$$0 \leq f_{13}^B \leq 45$$

$$0 \leq f_{21}^B \leq 45$$

$$0 \leq f_{23}^B \leq 45$$

$$0 \leq f_{31}^B \leq 45$$

$$0 \leq f_{32}^B \leq 45$$

$$0 \leq f_1^O \leq 45$$

$$0 \leq f_2^O \leq 45$$

$$0 \leq f_3^O \leq 45$$

$$100 \leq t_1^I \leq T_1^O - \Delta T_{min}$$

$$100 \leq t_2^I \leq T_2^O - \Delta T_{min}$$

$$100 \leq t_3^I \leq T_3^O - \Delta T_{min}$$

$$t_1^{O,min} \leq t_1^O \leq T_1^I - \Delta T_{min}$$
$$t_2^{O,min} \leq t_2^O \leq T_2^I - \Delta T_{min}$$
$$t_3^{O,min} \leq t_3^O \leq T_3^I - \Delta T_{min}$$

It should be noted that in this formulation the index i corresponds to the hot streams. T_i^I is the inlet temperature of hot stream i and T_i^O is the outlet temperature of hot stream i, as listed in Table 7.5.

7.3.2 Data

Stream data is given in Table 7.5, and match data are given in Table 7.6.

Table 7.5: Stream Data for Test Problem 2

STREAM	T in(K)	T out(K)	$FC_p(\frac{kW}{K})$
H1	210	130	25
H2	210	160	20
H3	210	180	50
C1	100	200	45

Table 7.6: Match Data for Test Problem 2

MATCH	Q_i (kW)	U_i ($\frac{kW}{m^2 K}$)
H1 C1	2000	0.5
H2 C1	1000	1.0
H3 C1	1500	2.0

Cost of Heat Exchangers = $1300A^{0.6}$

In addition to the stream and match data given in Tables 7.5 and 7.6, the model also requires the minimum temperatures approach, $\Delta T_{min} = 10°$, and the parameters f_i^{min}, the lower bound on the flowrate through exchanger i, and $t_i^{O,min}$, the minimum outlet temperature of the cold stream from exchanger i. These parameters are given in Table 7.7.

Table 7.7: Parameters for Example 2

i	f_i^{min}	$t_i^{O,min}$
1	20	144.444
2	10	122.222
3	45	133.333

7.3.3 Problem Statistics

This test problem is a nonlinear optimization problem with 27 variables and 19 equality constraints. Thirteen of the constraints are linear; the remaining six constraints are bilinear. The objective is a nonlinear but convex objective function.

7.3.4 Best Known Solution

The best known solution to this problem has been identified by [76]. The optimal network, shown in Figure 7.2, features a series piping configuration and an investment cost of $46,266. The variable levels are given in Tables 7.8 and 7.9.

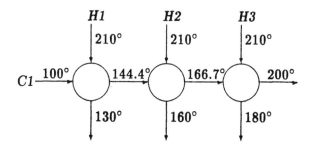

Figure 7.2: Best Known Solution to Test Problem 2

Table 7.8: Flowrate Variables at Best Known Solution

i	f_i^I	f_i^E	f_i^O	f_{ik}^B i \ k	1	2	3
1	45	45	0	1		0	0
2	0	45	0	2	45		0
3	0	45	45	3	0	45	

Table 7.9: Temperature Variable Levels for Test Problem 2

i	t_i^I	t_i^O	i	$\Delta T1_i$	$\Delta T2_i$
1	100	144.4	1	65.6	30
2	144.4	166.7	2	43.3	15.6
3	166.7	200	3	10	13.3

7.4 Test Problem 3 : 7SP4 Heat Exchanger Network Above the Pinch - NLP

Test problem 3 is taken from the literature problem 7SP4 ([180], [161]). This problem involves a designing a heat exchanger network for a system of one cold stream and six hot streams. The design problem has a pinch point at $430 - 410°F$ that partitions the network into two independent subproblems. Test problem 3 is a nonlinear optimization problem that identifies an optimal network configuration above the pinch.

7.4.1 Problem Formulation

$$minimize \quad 324\left[\frac{Q_1}{U_1(\frac{2}{3}(\Delta T1_1 \Delta T2_1)^{0.5} + \frac{1}{6}(\Delta T1_1 + \Delta T2_1))}\right]^{0.6} +$$

$$324\left[\frac{Q_2}{U_2(\frac{2}{3}(\Delta T1_2 \Delta T2_2)^{0.5} + \frac{1}{6}(\Delta T1_2 + \Delta T2_2))}\right]^{0.6} +$$

$$324\left[\frac{Q_3}{U_3(\frac{2}{3}(\Delta T1_3 \Delta T2_3)^{0.5} + \frac{1}{6}(\Delta T1_3 + \Delta T2_3))}\right]^{0.6}$$

subject to

$$f_1^I + f_2^I + f_3^I + f_4^I \quad = \quad 47$$

$$f_1^I + f_{12}^B + f_{13}^B + f_{14}^B - f_1^E \quad = \quad 0$$

$$f_2^I + f_{21}^B + f_{23}^B + f_{24}^B - f_2^E \quad = \quad 0$$

$$f_3^I + f_{31}^B + f_{32}^B + f_{34}^B - f_3^E \quad = \quad 0$$

$$f_4^I + f_{41}^B + f_{42}^B + f_{43}^B - f_4^E \quad = \quad 0$$

$$f_1^O + f_{21}^B + f_{31}^B + f_{41}^B - f_1^E \quad = \quad 0$$

$$f_2^O + f_{12}^B + f_{32}^B + f_{42}^B - f_2^E \quad = \quad 0$$

$$f_3^O + f_{13}^B + f_{23}^B + f_{43}^B - f_3^E \quad = \quad 0$$

$$f_4^O + f_{14}^B + f_{24}^B + f_{34}^B - f_4^E \quad = \quad 0$$

$$410 f_1^I + t_2^O f_{12}^B + t_3^O f_{13}^B + t_4^O f_{14}^B - t_1^I f_1^E \quad = \quad 0$$

$$410 f_2^I + t_1^O f_{21}^B + t_3^O f_{23}^B + t_4^O f_{24}^B - t_2^I f_2^E \quad = \quad 0$$

$$410 f_3^I + t_1^O f_{31}^B + t_2^O f_{32}^B + t_4^O f_{34}^B - t_3^I f_3^E \quad = \quad 0$$

$$410 f_4^I + t_1^O f_{41}^B + t_2^O f_{42}^B + t_3^O f_{43}^B - t_4^I f_4^E \quad = \quad 0$$

$$f_1^E (t_1^O - t_1^I) \quad = \quad Q_1$$

$$f_2^E (t_2^O - t_2^I) \quad = \quad Q_2$$

$$f_3^E (t_3^O - t_3^I) \quad = \quad Q_3$$

$$f_4^E (t_4^O - t_4^I) \quad = \quad Q_4$$

$$\Delta T1_1 \quad = \quad T_1^I - t_1^O$$

$$\Delta T2_1 \quad = \quad T_1^O - t_1^I$$

$$\Delta T1_2 \quad = \quad T_2^I - t_2^O$$

$$\Delta T2_2 \quad = \quad T_2^O - t_2^I$$

$$\Delta T1_3 \quad = \quad T_3^I - t_3^O$$

$$\Delta T2_3 \quad = \quad T_3^O - t_3^I$$

$$0 \le f_1^I \quad \le \quad 47$$

$$0 \le f_2^I \quad \le \quad 47$$

$$0 \le f_3^I \quad \le \quad 47$$

$$0 \le f_4^I \quad \le \quad 47$$

$$0 \leq f_{12}^B \leq 47$$

$$0 \leq f_{13}^B \leq 47$$

$$0 \leq f_{14}^B \leq 47$$

$$0 \leq f_{21}^B \leq 47$$

$$0 \leq f_{23}^B \leq 47$$

$$0 \leq f_{24}^B \leq 47$$

$$0 \leq f_{31}^B \leq 47$$

$$0 \leq f_{32}^B \leq 47$$

$$0 \leq f_{34}^B \leq 47$$

$$0 \leq f_1^O \leq 47$$

$$0 \leq f_2^O \leq 47$$

$$0 \leq f_3^O \leq 47$$

$$0 \leq f_4^O \leq 47$$

$$f_1^{min} \leq f_1^E \leq 47$$

$$f_2^{min} \leq f_2^E \leq 47$$

$$f_3^{min} \leq f_3^E \leq 47$$

$$f_4^{min} \leq f_4^E \leq 47$$

$$410 \leq t_1^I \leq T_1^O - \Delta T_{min}$$

$$410 \leq t_2^I \leq T_2^O - \Delta T_{min}$$

$$410 \leq t_3^I \leq T_3^O - \Delta T_{min}$$

$$410 \leq t_4^I \leq T_4^O - \Delta T_{min}$$

$$t_1^{min} \leq t_1^O \leq T_1^I - \Delta T_{min}$$

$$t_2^{min} \leq t_2^O \leq T_2^I - \Delta T_{min}$$

$$t_3^{min} \leq t_3^O \leq T_3^I - \Delta T_{min}$$

$$t_4^{min} \leq t_4^O \leq T_4^I - \Delta T_{min}$$

7.4.2 Data

The stream data is given in Table 7.10. Note that the parameters T_i^I and T_i^O correspond to the inlet and outlet temperatures of the hot streams. Match data is given in Table 7.11, and the lower bounds f_i^{min} and t_i^{min} are given in Table 7.12. Note that the minimum temperature approach, ΔT_{min}, equals $20°F$.

Table 7.10: Stream Data for Test Problem 3

STREAM	T in(F)	T out(F)	$FC_p(\frac{kBtu}{sF})$
H1	675	430	15
H2	590	450	11
H3	540	430	4.5
H4 (boiler)	801	800	
C1	410	710	47

Table 7.11: Match Data for Test Problem 3

MATCH	$Q_i \frac{Btu}{s}$	$U_i (\frac{Btu}{ft^2F})$
H1 C1	3675	0.1
H2 C1	1540	0.07
H3 C1	495	0.06
H4 C1	8390	

Table 7.12: Flowrate and temperature lower bounds for Test Problem 3

i	f_i^{min}	t_i^{min}
1	15	488.191
2	9.625	442
3	4.5	420.532
4	22.676	588.511

7.4.3 Problem Statistics

This problem involves a system of 38 variables and 23 equality constraints. Of these constraints, 15 are linear constraints and the remaining 8 constraints are bilinear. The objective function is a nonlinear but convex function.

7.4.4 Best Known Solution

The best known solution to this problem involves piping the cold stream in a series-parallel configuration. The optimum is at \$34,633. This solution is shown in Figure 7.3, and the variable levels are given in Table 7.13.

Table 7.13: Flowrate and Temperature Variables at the Best Known Solution of Test Problem 3

i	f_i^I	f_{i1}^B	f_{i2}^B	f_{i3}^B	f_i^E	f_i^O	t_i^I	t_i^O	$\Delta T1_i$
1	24.348				24.348		410	560.933	114.067
2	14.882				14.882		410	513.482	76.518
3	7.770				7.770		410	473.709	66.291
4		24.348	14.882	7.770	47	47	531.489	710	

7.5 Test Problem 4 : 7SP4 Heat Exchanger Network Below the Pinch - NLP

Test problem 4 is taken from the literature problem 7SP4 ([180], [161]). This problem involves a designing a heat exchanger network for a system of one cold stream and six hot streams, with extra cooling provided by cooling water and extra heating provided by a boiler. The design problem has a pinch point at $430 - 410°F$ that partitions the network into two independent subproblems. Test problem 4 is a nonlinear programming (NLP) optimization problem that identifies the network configuration for the subnetwork below the pinch point without considering the match between hot stream H5 and cooling water CW.

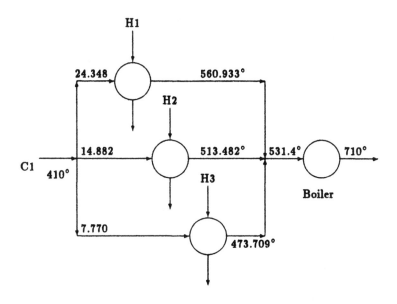

Figure 7.3: Best Known Solution to Test Problem 3

7.5.1 Problem Formulation

$$
\begin{aligned}
min \quad & 324\left\{\frac{Q_{11}}{U_{11}(\frac{2}{3}(\Delta T1_{11}\Delta T2_{11})^{0.5} + \frac{1}{6}(\Delta T1_{11} + \Delta T2_{11}))}\right\}^{0.6} + \\
& 324\left\{\frac{Q_{1W}}{U_{1W}(\frac{2}{3}(\Delta T1_{1W}\Delta T2_{1W})^{0.5} + \frac{1}{6}(\Delta T1_{1W} + \Delta T2_{1W}))}\right\}^{0.6} + \\
& 324\left\{\frac{Q_{31}}{U_{31}(\frac{2}{3}(\Delta T1_{31}\Delta T2_{31})^{0.5} + \frac{1}{6}(\Delta T1_{31} + \Delta T2_{31}))}\right\}^{0.6} + \\
& 324\left\{\frac{Q_{41}}{U_{41}(\frac{2}{3}(\Delta T1_{41}\Delta T2_{41})^{0.5} + \frac{1}{6}(\Delta T1_{41} + \Delta T2_{41}))}\right\}^{0.6} + \\
& 324\left\{\frac{Q_{61}}{U_{61}(\frac{2}{3}(\Delta T1_{61}\Delta T2_{61})^{0.5} + \frac{1}{6}(\Delta T1_{61} + \Delta T2_{61}))}\right\}^{0.6}
\end{aligned}
$$

subject to

$$
\begin{aligned}
f_1^{CI} + f_3^{CI} + f_4^{CI} + f_6^{CI} &= 47 \\
f_1^{CI} + f_{13}^{CB} + f_{14}^{CB} + f_{16}^{CB} - f_1^{CE} &= 0 \\
f_3^{CI} + f_{31}^{CB} + f_{34}^{CB} + f_{36}^{CB} - f_3^{CE} &= 0 \\
f_4^{CI} + f_{41}^{CB} + f_{43}^{CB} + f_{46}^{CB} - f_4^{CE} &= 0 \\
f_6^{CI} + f_{61}^{CB} + f_{63}^{CB} + f_{64}^{CB} - f_6^{CE} &= 0 \\
f_1^{CO} + f_{31}^{CB} + f_{41}^{CB} + f_{61}^{CB} - f_1^{CE} &= 0 \\
f_3^{CO} + f_{13}^{CB} + f_{43}^{CB} + f_{63}^{CB} - f_3^{CE} &= 0 \\
f_4^{CO} + f_{14}^{CB} + f_{34}^{CB} + f_{64}^{CB} - f_4^{CE} &= 0 \\
f_6^{CO} + f_{16}^{CB} + f_{36}^{CB} + f_{46}^{CB} - f_6^{CE} &= 0 \\
60f_1^{CI} + t_3^{CO}f_{13}^{CB} + t_4^{CO}f_{14}^{CB} + t_6^{CO}f_{16}^{CB} - t_1^{CI}f_1^{CE} &= 0 \\
60f_3^{CI} + t_1^{CO}f_{31}^{CB} + t_4^{CO}f_{34}^{CB} + t_6^{CO}f_{36}^{CB} - t_3^{CI}f_3^{CE} &= 0 \\
60f_4^{CI} + t_1^{CO}f_{41}^{CB} + t_3^{CO}f_{43}^{CB} + t_6^{CO}f_{46}^{CB} - t_4^{CI}f_4^{CE} &= 0 \\
60f_6^{CI} + t_1^{CO}f_{61}^{CB} + t_3^{CO}f_{63}^{CB} + t_4^{CO}f_{64}^{CB} - t_6^{CI}f_6^{CE} &= 0 \\
f_1^{CE}(t_1^{CO} - t_1^{CI}) &= Q_{11} \\
f_3^{CE}(t_3^{CO} - t_3^{CI}) &= Q_{31} \\
f_4^{CE}(t_4^{CO} - t_4^{CI}) &= Q_{41}
\end{aligned}
$$

$$f_6^{CE}(t_6^{CO} - t_6^{CI}) = Q_{61}$$

$$f_1^{HI} + f_W^{HI} = 15$$

$$f_1^{HI} + f_{1W}^{HB} - f_1^{HE} = 0$$

$$f_W^{HI} + f_{W1}^{HB} - f_W^{HE} = 0$$

$$f_1^{HO} + f_{W1}^{HB} - f_1^{HE} = 0$$

$$f_W^{HO} + f_{1W}^{HB} - f_W^{HE} = 0$$

$$430 f_1^{HI} + t_W^{HO} f_{1W}^{HB} - t_1^{HI} f_1^{HE} = 0$$

$$430 f_W^{HI} + t_1^{HO} f_{W1}^{HB} - t_W^{HI} f_W^{HE} = 0$$

$$f_1^{HE}(t_1^{HI} - t_1^{HO}) = Q_{11}$$

$$f_W^{HE}(t_W^{HI} - t_W^{HO}) = Q_{1W}$$

$$\Delta T1_{11} = t_1^{HI} - t_1^{CO}$$

$$\Delta T2_{11} = t_1^{HO} - t_1^{CI}$$

$$\Delta T1_{1W} = t_W^{HI} - T_W^{CO}$$

$$\Delta T2_{1W} = t_W^{HO} - T_W^{CI}$$

$$\Delta T1_{31} = T_3^{HI} - t_3^{CO}$$

$$\Delta T2_{31} = T_3^{HO} - t_3^{CI}$$

$$\Delta T1_{41} = t_4^{HI} - t_4^{CO}$$

$$\Delta T2_{41} = t_4^{HO} - t_4^{CI}$$

$$\Delta T1_{61} = t_6^{HI} - t_6^{CO}$$

$$\Delta T2_{61} = t_6^{HO} - t_6^{CI}$$

$$\Delta T1_{11} \geq \Delta T_{min}$$

$$\Delta T1_{1W} \geq \Delta T_{min}$$

$$\Delta T1_{31} \geq \Delta T_{min}$$

$$\Delta T1_{41} \geq \Delta T_{min}$$

$$\Delta T1_{61} \geq \Delta T_{min}$$

$$\Delta T2_{11} \geq \Delta T_{min}$$

$$\Delta T2_{1W} \geq \Delta T_{min}$$

$$\Delta T2_{31} \geq \Delta T_{min}$$

$$\Delta T2_{41} \geq \Delta T_{min}$$

$$\Delta T2_{61} \geq \Delta T_{min}$$

$$0 \leq f_1^{CI} \leq 47$$

$$0 \leq f_3^{CI} \leq 47$$

$$0 \leq f_4^{CI} \leq 47$$

$$0 \leq f_6^{CI} \leq 47$$

$$0 \leq f_1^{CO} \leq 47$$

$$0 \leq f_3^{CO} \leq 47$$

$$0 \leq f_4^{CO} \leq 47$$

$$0 \leq f_6^{CO} \leq 47$$

$$0 \leq f_{13}^{CB} \leq 47$$

$$0 \leq f_{14}^{CB} \leq 47$$

$$0 \leq f_{16}^{CB} \leq 47$$

$$0 \leq f_{31}^{CB} \leq 47$$

$$0 \leq f_{34}^{CB} \leq 47$$

$$0 \leq f_{36}^{CB} \leq 47$$

$$0 \leq f_{41}^{CB} \leq 47$$

$$0 \leq f_{43}^{CB} \leq 47$$

$$0 \leq f_{46}^{CB} \leq 47$$

$$0 \leq f_{61}^{CB} \leq 47$$

$$0 \leq f_{63}^{CB} \leq 47$$

$$0 \leq f_{64}^{CB} \leq 47$$

$$f_1^{Cmin} \leq f_1^{CE} \leq 47$$

$$f_3^{Cmin} \leq f_3^{CE} \leq 47$$

$$f_4^{Cmin} \leq f_4^{CE} \leq 47$$

$$f_6^{Cmin} \leq f_6^{CE} \leq 47$$

$$0 \le f_1^{HI} \le 15$$

$$0 \le f_W^{HI} \le 15$$

$$0 \le f_{1W}^{HB} \le 15$$

$$0 \le f_{W1}^{HB} \le 15$$

$$0 \le f_1^{HO} \le 15$$

$$0 \le f_W^{HO} \le 15$$

$$f_1^{Hmin} \le f_1^{HE} \le 15$$

$$f_W^{Hmin} \le f_W^{HE} \le 15$$

$$60 \le t_1^{CI} \le 410$$

$$60 \le t_3^{CI} \le 410$$

$$60 \le t_4^{CI} \le 410$$

$$60 \le t_6^{CI} \le 410$$

$$60 \le t_1^{CO} \le 410$$

$$60 \le t_3^{CO} \le 410$$

$$60 \le t_4^{CO} \le 410$$

$$60 \le t_6^{CO} \le 410$$

$$80 \le t_1^{HI} \le 430$$

$$100 \le t_W^{HI} \le 430$$

$$80 \le t_1^{HO} \le 430$$

$$100 \le t_W^{HO} \le 430$$

7.5.2 Data

The data for this problem consists of stream, matches and variable bounds. The stream data is given in Table 7.14, and match data is given in Table 7.5.2. Lower bounds on flowrates through exchangers are given in Table 7.5.2.

Table 7.14: Stream Data for Test Problem 4

STREAM	T in(F)	T out(F)	$FC_p(\frac{kBtU}{hrF})$
H1	430	150	15
H3	430	115	4.5
H4	430	345	60
H6	300	230	125
C1	60	410	47
CW	80	140	11.03

Table 7.15: Match Data for Test Problem 4

MATCH	Q(kBtU/hr)	U $(\frac{kBtU}{hrft^2F})$
H1 C1	1182.5	0.1
H1 CW	3017.5	0.08
H3 C1	1417.5	0.06
H4 C1	5100.0	0.07
H6 C1	8750.0	0.055

Table 7.16: Lower Bounds on Flowrates through Exchangers

i	f_i^{Cmin}	j	f_i^{Hmin}
1	3.379	1	3.379
3	4.050	2	9.144
4	14.571		
6	39.773		

7.5.3 Problem Statistics

This problem contains 54 variables and 36 constraints. 26 constraints are linear, and the remaining 10 constraints are bilinear. The objective is a nonlinear but convex function. Upper and lower bounds have been provided for each variable.

7.5.4 Best Known Solution

The best known solution is reported in [76]. It involves arranging the matches of hot stream 1 in a series configuration, and the matches of cold stream 1 in a series-parallel configuration. This network is shown in Figure 7.4. The value of the objective function for this solution is \$91,142. The levels of the variables at this point are listed in Tables 7.17.

Table 7.17: Cold Stream 1 Flowrate Variable Levels

i	f_i^{CI}	f_i^{CE}	f_i^{CO}	f_{i1}^{CB}	f_{i3}^{CB}	f_{i4}^{CB}	f_{i6}^{CB}
1		41.45					41.45
3	5.55	5.55					
4		47	47	41.45	5.55		
6	41.45	41.45					

Table 7.18: Cold Stream 1 Temperature Variable Levels

i	t_i^{CI}	t_i^{CO}
1	271.1	299.6
3	60	271.1
4	301.5	410
6	60	315.4

Table 7.19: Hot Stream 1 Flowrate and Temperature Variable Levels

j	f_j^{HI}	f_j^{HE}	f_j^{HO}	f_{j1}^{HB}	f_{j2}^{HB}	t_j^{HI}	t_j^{HO}
1	15	15				430	351.2
2		15	15	15		351.2	150

Table 7.20: Temperature Differences

(ij)	$\Delta T1_{ij}$	$\Delta T2_{ij}$
(11)	130.4	80.1
(12)	211.2	70
(31)	114.6	55
(41)	20	43.5
(61)	28.9	170

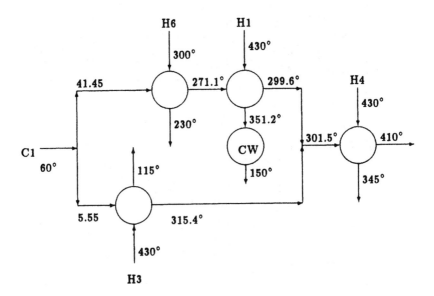

Figure 7.4: Best Known Solution to Test Problem 4

7.6 Test Problem 5 : Heat Exchanger Network Optimization - (MINLP)

This problem involves finding the optimal set of matches, heat load distribution, and network configuration for a system of 3 hot streams, 3 cold streams, and one cold utility (cooling water). This problem is taken from [76]. The mathematical optimization model is a mixed integer nonlinear programming (MINLP) model as it contains both integer and continuous variables, and involves a nonlinear objective function and constraint set. The constraint set is composed of two subsets; (a) the transshipment model of heat flow and (b) the hyperstructure model for simultaneous match and network configuration optimization ([76]). It should be noted that the objective function minimizes the investment cost of the heat exchanger network.

7.6.1 Problem Formulation

$$min \sum_{(ij)\in M(ij)} 1300 A_{ij}^{0.6} Y_{ij}$$

$$subject\ to$$

$$\sum_{i\in HCT(ijt)} q_{ijt} = Q_{jt}^C \quad j \in C; \quad t \in T$$

$$\sum_{j\in HCT(ijt)} q_{ijt} + R_{it} - R_{i-1,t} = Q_{it}^H \quad i \in H; \quad t \in T$$

$$\sum_{t\in HCT(ijt)} q_{ijt} = Q_{ij} \quad (ij) \in M(ij)$$

$$Q_{ij} - U_{ij}^Q Y_{ij} \leq 0 \quad (ij) \in M(ij)$$

$$\sum_{(ij)\in M(ij)} Y_{ij} \leq 6$$

$$\sum_{j\in M(ij)} f_{ij}^{HI} = F_i^{HCP} \quad i \in H$$

$$f_{ij}^{HI} + \sum_{m\in C;\ m\neq j} f_{ijm}^{HB} = f_{ij}^{HE} \quad (ij) \in M(ij)$$

$$f_{ij}^{HO} + \sum_{m \in C;\ m \neq j} f_{imj}^{HB} = f_{ij}^{HE} \quad (ij) \in M(ij)$$

$$T_i^{HI} f_{ij}^{HI} + \sum_{m \in C;\ m \neq j} t_{im}^{HO} f_{ijm}^{HB} = t_{ij}^{HI} f_{ij}^{HE} \quad (ij) \in M(ij)$$

$$f_{ij}^{HE}(t_{ij}^{HI} - t_{ij}^{HO}) = Q_{ij} \quad (ij) \in M(ij)$$

$$f_{ij}^{HE} - F_i^{HCP} Y_{ij} \leq 0 \quad (ij) \in M(ij)$$

$$\Delta T_{ij}^{max} f_{ij}^{HE} \geq Q_{ij} \quad (ij) \in M(ij)$$

$$\sum_{i \in M(ij)} f_{ij}^{CI} = F_j^{CCP} \quad j \in C$$

$$f_{ij}^{CI} + \sum_{k \in H;\ k \neq i} f_{ikj}^{CB} = f_{ij}^{CE} \quad (ij) \in M(ij)$$

$$f_{ij}^{CO} + \sum_{k \in H;\ k \neq i} f_{kij}^{CB} = f_{ij}^{CE} \quad (ij) \in M(ij)$$

$$T_j^{CI} f_{ij}^{CI} + \sum_{k \in H;\ k \neq i} t_{kj}^{CO} f_{ikj}^{CB} = t_{ij}^{CI} f_{ij}^{CE} \quad (ij) \in M(ij)$$

$$f_{ij}^{CE}(t_{ij}^{CO} - t_{ij}^{CI}) = Q_{ij} \quad (ij) \in M(ij)$$

$$f_{ij}^{CE} - F_j^{CCP} Y_{ij} \leq 0 \quad (ij) \in M(ij)$$

$$\Delta T_{ij}^{max} f_{ij}^{CE} \geq Q_{ij} \quad (ij) \in M(ij)$$

$$\Delta T_{ij}^1 = t_{ij}^{HI} - t_{ij}^{CO} \quad (ij) \in M(ij)$$

$$\Delta T_{ij}^2 = t_{ij}^{HO} - t_{ij}^{CI} \quad (ij) \in M(ij)$$

$$LMTD_{ij} = 2/3(\Delta T_{ij}^1 \Delta T_{ij}^2)^{1/2} + 1/6(\Delta T_{ij}^1 + \Delta T_{ij}^2) \quad (ij) \in M(ij)$$

$$A_{ij} = \frac{Q_{ij}}{0.8 LMTD_{ij}} \quad (ij) \in M(ij)$$

$$q_{ijt} \geq 0 \quad (ijt) \in HCT(ijt)$$

$$Q_{ij} \geq 0 \quad (ij) \in M(ij)$$

$$R_{it} \geq 0 \quad i \in H;\ t \in T$$

$$0 \leq f_{ij}^{HI} \leq F_i^{HCP} \quad (ij) \in M(ij)$$

$$0 \leq f_{ijm}^{HB} \leq F_i^{HCP} \quad (ij) \in M(ij); (im) \in M(ij); m \neq j$$

$$0 \leq f_{ij}^{HO} \leq F_i^{HCP} \quad (ij) \in M(ij)$$

$$0 \leq f_{ij}^{CI} \leq F_j^{CCP} \quad (ij) \in M(ij)$$

$$0 \leq f_{ikj}^{CB} \leq F_j^{CCP} \quad (ij) \in M(ij); (kj) \in M(kj); k \neq i$$

$$0 \leq f_{ij}^{CO} \leq F_j^{CCP} \quad (ij) \in M(ij)$$
$$\Delta T1_{ij}, \ \Delta T2_{ij}, \ LMTD_{ij} \geq \Delta T_{min} \quad (ij) \in M(ij)$$
$$Y_{ij} \in [0,1] \quad (ij) \in M(ij)$$

7.6.2 Data

The data provided for this problem consists of (a) sets and (b) parameters. The sets are as follows:

$i \in H$ The set of hot streams i. Note that there are three hot streams in this problem, and so the index $i = 1, ..., 3$.

$j \in C$ The set of cold streams and cold utilities j. Note that there are three cold process streams, as well as cooling water, and so the index $j = 1, ..., 4$. The index for cooling water is $j = 4$. Note that cooling water can be modelled as a regular stream in this problem.

$t \in T$ The set of temperature intervals in the transshipment model. This set is generated by partitioning the temperature range into a set of intervals with the inlet temperatures of the streams. In this problem $t = 1, ..., 4$.

$(ij) \in M(ij)$ The set of allowable matches between hot and cold process streams. Note that the first index of the set denotes the hot stream of the potential match, while the second index denotes the cold stream of the potential match. There are no restrictions on stream matches in this problem, and so $M(ij) = \{i = 1, ..., 3; \ j = 1, ..., 4\}$.

$HCT(ijt)$ The set of temperature intervals t where cold stream j can absorb heat from hot stream i. This set is determined by the following rule: $H(ijt) = \{(ijt) \mid j \in t \text{ and } i \in t' t' \leq t\}$. Set $HCT(ijt)$ is also shown graphically in Figure 7.5.

In addition to the sets, several parameters are also required. These include the minimum temperature approach, ΔT_{min}, which equals $10° K$, as well as the

(ij)	$t = 1$	$t = 2$	$t = 3$	$t = 4$
(11)	x	x	x	x
(12)		x	x	
(13)		x	x	
(14)				x
(21)		x	x	x
(22)		x	x	
(23)		x	x	
(24)				x
(31)			x	x
(32)			x	
(33)			x	
(34)				x

Figure 7.5: Set $HCT(ijt)$ for Test Problem 5

stream data (shown in Table 7.21), the heat duties of the streams within the transshipment model, Q_{it}^H and Q_{jt}^C, given in Tables 7.22 and 7.23, and the maximum temperature changes for the streams, ΔT_{ij}^{max}, is given in Table 7.24. It should be noted that the parameters T_i^H and T_j^C refer to the inlet temperatures T in(K) of hot stream i and cold stream j, while the parameters F_i^{HCP} and F_j^{CCP} refer to the heat capacity flowrates $FC_p(\frac{kW}{K})$ of hot stream i and cold stream j, as listed in Table 7.21.

Table 7.21: Stream Data for Test Problem 5

STREAM	T in(K)	T out(K)	$FC_p(\frac{kW}{K})$
H1	500	350	10
H2	450	350	12
H3	400	320	8
C1	300	480	9
C2	340	420	10
C3	340	400	8
CW	300	320	22

Table 7.22: Parameter Q_{it}^H in Test Problem 5

i	$t = 1$	$t = 2$	$t = 3$	$t = 4$
1	500	500	500	
2		600	600	
3			400	240

Table 7.23: Parameter Q_{jt}^C in Test Problem 5

j	$t = 1$	$t = 2$	$t = 3$	$t = 4$
1	360	450	450	360
2		300	500	
3		80	400	
4				440

Table 7.24: Parameter ΔT_{ij}^{max} in Test Problem 5

i	$j = 1$	$j = 2$	$j = 3$	$j = 4$
1	195	155	155	195
2	145	105	105	145
3	95	55	55	95

7.6.3 Problem Statistics

This problem contains 268 variables and 242 constraints, excluding variable bounds. Of the 268 variables, 12 are integer variables taking $[0,1]$ values only. The remaining 256 are continuous variables. Of the 242 constraints, 170 are linear, 48 are bilinear, and the remaining 24 are more complexly nonlinear. The integer variables appear linearly in 36 linear constraints, and nonlinearly in the objective function. The objective function is a nonlinear, nonconvex function involving both continuous and integer variables.

7.6.4 Best Known Solution

The best known solution to this problem had a value of $49,352. This solution is shown in Figure 7.6, and selected variable levels are reported in Table 7.6.4.

Table 7.25: Values of Selected Variables for Test Problem 5

$Y_{ij} = 1$	Q_{ij}	A_{ij}
(11)	1500	68.2
(21)	120	3.1
(22)	800	41.8
(23)	280	12.3
(33)	200	11.1
(34)	440	15.9

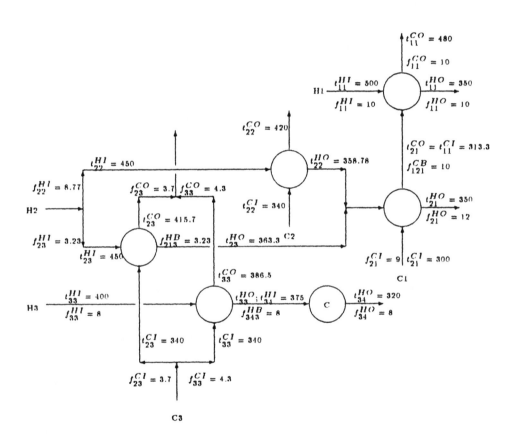

Figure 7.6: Best Known Solution to Test Problem 5

Chapter 8

Phase and Chemical Reaction Equilibrium test problems

8.1 Problem Statement

At the root of most chemical process design problems such as distillation column design or alternative separation systems and reactor design, lies the fundamental problem of phase and chemical equilibrium. This problem can be stated as:

> *Given a set of feed conditions for a reacting or non-reacting multicomponent system at a given temperature and pressure, determine the number and state of phases existing at equilibrium as well as the composition and quantity of each.*

Optimization approaches for the phase and chemical equilibrium problem began with the pioneering work of [242]. Selective pieces of research work are described in [221], [148], [149], and [177]. The classical mathematical formulation of phase and chemical equilibrium problems corresponds to a non-convex NLP model which consists of a nonlinear objective function (Gibbs free energy) subject to a linear set os constraints (mass and/or component balances). The mathematical properties of the nonlinear objective function are completely dependent on the mathematical form of the equation of state (EOS) and/or fugacity correlations chosen to represent each of the phases that may exist at equilibrium.

These functionalities can be relatively simple for the case of an ideal vapor state or highly complex for systems in which the **NRTL** (Non-Random Two-Liquid model) or **UNIQUAC** (UNIversal QUAsi-Chemical model) or EOS near the critical region are used to represent nonideal liquid-liquid interactions.

The nonconvexities of the formulation arise due to the form of the objective function and can lead to multiple meta-stable equilibria points or points whose multiphase solution converges to a trivial solution. Trivial solutions can correspond to a saddle point solution if the feed composition lies under the spinodal curve or a local minimum if the feed is between the spinodal and binodal curves. The globally optimal solution of the mathematical formulation corresponds to the *true* equilibrium solution (i.e. the one found in nature).

8.2 Test Problem 1 : Ethanol-Acetic Acid Reaction L-V Equilibrium

This test problem is a modification of example 2 of [177] and involves an equimolar mixture of ethanol and acetic acid reacts reversibly according to the reaction :

$$\text{EtOH} + \text{HAc} = \text{EtAc} + \text{H}_2\text{O}$$

Simultaneous chemical and phase equilibria can lead to one or two phase solutions (vapor only / liquid only / vapor-liquid). For this system, the Wilson equation has been used to model the liquid phase and, the vapor phase is treated as ideal. Conditions of 358K and 1 atm have been chosen for the study of simultaneous phase and chemical equilibrium. A nonlinear optimization problem with linear constraints and a nonconvex nonlinear objective function is formulated.

8.2.1 Problem Formulation

$$\min G =$$
$$\sum_{i \in I_c} n_i^{vap} \left[\Delta G_i^{vap^0} / RT + \left\{ \ln(n_i^{vap}) - \ln\left(\sum_j n_j^{vap}\right) \right\} \right]$$

$$+ \sum_{i \in I_c} n_i^{liq} \left(\Delta G_i^{liq^0}/RT + \left\{ 1 - \ln \left(\sum_j n_j^{liq} \Lambda_{ij} \right) - \sum_j \left\{ \frac{n_j^{liq} \Lambda_{ji}}{\sum_l n_l^{liq} \Lambda_{jl}} \right\} \right\} + (\ln n_i^{liq} \right)$$

subject to

$$\sum_{i \in I_c} \sum_{k \in I_p} a_{ei} n_i^k = b_e \qquad e \in I_e$$

$$n_i^k \geq 0 \qquad i \in I_c, \; k \in I_p$$

$$(i, j, l) \in I_c = \{EtOH, \; HAc, \; EtAc, \; H2O\}$$

$$k \in I_p = \{Vap, \; Liq\}$$

$$e \in I_e = \{C, \; H, \; O\}$$

8.2.2   Data

Parameter Λ_{ij}				
	EtOH	**HAc**	**EtAc**	**H2O**
EtOH	1.000	2.282	0.767	0.153
HAc	0.276	1.000	0.618	0.268
EtAc	0.550	0.893	1.000	0.123
H2O	0.920	1.226	0.149	1.000

Parameter $\Delta G_i^{k^0}$		
	Vap	**Liq**
EtOH	-37.111	-36.764
HAc	-87.326	-88.596
EtAc	-72.848	-72.637
H2O	-54.050	-54.453

Parameter a_{ei}				
	EtOH	**HAc**	**EtAc**	**H2O**
C	2.000	2.000	4.000	0.000
H	6.000	4.000	8.000	2.000
O	1.000	2.000	2.000	1.000

Parameter b_e		
$b(C)$	=	2.0
$b(H)$	=	5.0
$b(O)$	=	1.5

Other Data

Parameter $RT = 0.711$

8.2.3 Problem Statistics

Number of constraints = 4
Number of variables = 9
Implicit lower bounds = 9

8.2.4 Best Known Solution

Solution		
$G = -90.0521$		
	Vap	Liq
n_{EtOH}	0.086	0.007
n_{HAc}	0.047	0.046
n_{EtAc}	0.388	0.019
n_{H2O}	0.335	0.071

8.3 Test Problem 2 : Toluene-Water-Analine L-L Equilibrium

This example, which is a modified version of example 3 of [177], involves a liquid-liquid non-reacting system with three components. It is desired to find the equilibrium concentrations of the three components and whether the system is one or two liquid phases at equilibrium. This problem is treated as a nonconvex nonlinear programming problem. This problem exhibits strong trivial solution behavior in that two phases may be postulated but a homogeneous solution can be obtained in which both phases contain a liquid of identical composition and thus is equivalent to a single phase metastable solution. Convergence to the trivial solution is highly dependent on the starting point.

8.3.1 Problem Formulation

$$\min G = $$

$$\sum_{i \in I_c} \sum_{k \in I_p} n_i^k \left(\Delta G_i^{k^0} / RT \; + \; \left\{ \frac{\sum_j \tau_{ji}^k G_{ji}^k n_j^k}{\sum_l G_{li}^k n_l^k} + \sum_j \frac{G_{ij}^k n_j^k}{\sum_l G_{lj}^k n_l^k} \left\{ \tau_{ij}^k - \frac{\sum_l \tau_{lj}^k G_{lj}^k n_l^k}{\sum_l G_{lj}^k n_l^k} \right\} \right\} \right.$$

$$\left. + \; \ln n_i^k - \ln \left(\sum_{i \in I_c} n_i^k \right) \right)$$

subject to

$$\sum_{k \in I_p} n_i^k \;\; = \;\; n_i^T \quad i \in I_c$$

$$n_i^k \;\; \geq \;\; 0 \quad i \in I_c \; k \in I_p$$

$$(i, j, l) \in I_c \;\; = \;\; \{C7H8, \; H2O, \; Analine\}$$

$$k \in I_p \;\; = \;\; \{Liq1, \; Liq2\}$$

8.3.2 Data

Parameter G_{ij}			
	C7H8	H2O	Analine
C7H8	1.000	0.294	0.619
H2O	0.145	1.000	0.240
Analine	0.990	0.646	1.000

Parameter τ_{ij}			
	C7H8	**H2O**	**Analine**
C7H8	0.000	4.930	1.598
H2O	7.771	0.000	4.185
Analine	0.035	1.279	0.000

Parameter $\Delta G_i^{k^\circ}$		
	Liq1	**Liq2**
C7H8	27.190	27.190
H2O	-56.702	-56.702
Analine	35.630	35.630

Parameter n_i^T		
n_{C7H8}^T	=	0.2995
n_{H2O}^T	=	0.1998
$n_{Analine}^T$	=	0.4994

Other Data

Parameter $RT = 0.592$

8.3.3 Problem Statistics

Number of constraints = 13

Number of variables = 10

Implicit lower bounds = 10

8.3.4 Best Known Solution

Solution		
$G = 24.3412$		
	Liq1	Liq2
n_{C7H8}	0.188	0.111
n_{H2O}	0.126	0.074
$n_{Analine}$	0.314	0.186

8.4 Test Problem 3 : Benzene-Acetonotrile-Water L-L-V Equilibrium

This example, which corresponds to example 4 of [177] with the two liquid phase solution, involves a three component non-reacting phase equilibria problem. The liquid phase is modeled via the **NRTL** model. The vapor phase is treated as ideal. At a set of intermediate conditions, (333K 0.769 atm), a three phase solution is obtained. If the pressure is increased (333K, 1 atm) the vapor phase is eliminated and a two-phase liquid-liquid system results. Finally if the temperature and pressure are decreased (300K, 0.1 atm), a two-phase system with a vapor phase and one liquid phase is the result. This system exhibits a number of local solutions as well as a strong homogenous solution for the two-liquid phase systems.

8.4.1 Problem Formulation

$$\min G =$$

$$\sum_{i \in I_c} n_i^{vap} \left(\Delta G_i^{vap^0} / RT \ + \ \left\{ \ln \left(n_i^{vap} \right) - \ln \left(\sum_j n_j^{vap} \right) \right\} \right.$$

$$+ \sum_{i \in I_c} \sum_{k \in I_{lq}} n_i^k \left(\Delta G_i^{k^0} / RT \ + \ \left\{ \frac{\sum_j \tau_{ji}^k G_{ji}^k n_j^k}{\sum_l G_{li}^k n_l^k} + \sum_j \frac{G_{ij}^k n_j^k}{\sum_l G_{lj}^k n_l^k} \left\{ \tau_{ij}^k - \frac{\sum_l \tau_{lj}^k G_{lj}^k n_l^k}{\sum_l G_{lj}^k n_l^k} \right\} \right\} \right.$$

$$+ \ \ln n_i^k - \ln \left(\sum_{i \in I_c} n_i^k \right) \Big)$$

subject to

$$\sum_{k \in I_p} n_i^k = n_i^T \qquad i \in I_c$$

$$n_i^k \geq 0 \qquad i \in I_c,\ k \in I_p$$

$$(i,j,l) \in I_c = \{C6H6,\ CH3CN,\ H2O\}$$
$$k \in I_p = \{Vap,\ Liq1,\ Liq2\}$$
$$k \in I_{lq} = \{Liq1,\ Liq2\}$$

8.4.2 Data

Parameter $\mathcal{G}_{ij}$			
	C6H6	CH3CN	H2O
C6H6	1.000	0.413	0.383
CH3CN	0.943	1.000	0.878
H2O	0.386	0.638	1.000

Parameter τ_{ij}			
	C6H6	CH3CN	H2O
C6H6	0.000	0.998	3.883
CH3CN	0.066	0.000	0.364
H2O	3.850	1.262	0.000

Parameter $\Delta G_i^{k^o}$			
	Vap	Liq1	Liq2
C6H6	32.236	31.793	31.793
CH3CN	25.367	24.889	24.889
H2O	-54.285	-55.366	-55.366

Parameter n_i^T		
n_{C6H6}^T	=	0.3436
n_{CH3CN}^T	=	0.3092
n_{H2O}^T	=	0.3472

Other Data

Parameter $RT = 0.662$

8.4.3 Problem Statistics

Number of constraints = 10

Number of variables = 7

Implicit lower bounds = 7

8.4.4 Best Known Solution

	Solution		
	$G = -1.3523$		
	Vap	**Liq1**	**Liq2**
n_{C6H6}	0.0	0.336	0.008
n_{CH3CN}	0.0	0.245	0.064
n_{H2O}	0.0	0.050	0.298

Chapter 9

Complex Chemical Reactor Network test problems

9.1 Problem Statement

The reactor network optimization problem can be stated as follows:

> *Given a reaction mechanism and the kinetics that describe it, determine a reactor network that optimizes a prescribed performance index. The considered performance index will be in general a function of the outlet stream compositions and the reactor volumes. The resulting optimal reactor network should provide information about :*
>
> *(i) the number of reactors ;*
>
> *(ii) the type of reactors ;*
>
> *(iii) the volumes of the reactors ;*
>
> *(iv) the appropriate feeding, recycling and bypassing strategy ; and*
>
> *(v) the configuration of the reactor network (i.e. interconnection of the reactor units) and the optimal values of the flowrates and compositions.*

Optimization based approaches for complex reactor networks have resulted in nonconvex nonlinear programming NLP models ([107]; [116]; [2]; [125]) and nonconvex mixed-integer nonlinear programming MINLP models ([125]). The objective function can be linear or nonlinear. The constraints represent a set of nonlinear expressions in terms of the continuous variables. Although bilinearities will always appear in the mass balances of the mixers, the mass balances around each reactor unit will result in nonlinearities that depend on the specific kinetics of the problem. Thus, bilinearities will appear for first order kinetics, trilinearities for second order kinetics and so on. Because the set of constraints is of this type, the feasible region of the problem is a nonconvex set.

9.2 Test Problem 1 : Van der Vusse reaction

This test problem is taken from [125] and involves the Van der Vusse reaction :

- *reaction mechanism:*

$$A \xrightarrow{k_1} B \quad (first\ order\ reaction\ ) \tag{9.1}$$
$$B \xrightarrow{k_2} C \quad (first\ order\ reaction\ ) \tag{9.2}$$
$$2\ A \xrightarrow{k_3} D \quad (second\ order\ reaction\ ) \tag{9.3}$$

- *objective function:* Maximization of yield of B.

9.2.1 Problem Formulation

$$\text{MAX} \quad x_{5,B}$$

$$F_1 - F_2 - F_3 = 0$$
$$F_4 + F_7 - F_9 = 0$$
$$F_5 + F_6 + F_8 - F_{10} = 0$$
$$F_7 + F_8 - F_{23} - F_{12} = 0$$
$$F_7 \cdot x_{3,A} + F_8 \cdot x_{4,A} - F_{11} \cdot x_{5,A} - F_{12} \cdot x_{5,A} = 0$$

$$F_7 \cdot x_{3,B} + F_8 \cdot x_{4,B} - F_{11} \cdot x_{5,B} - F_{12} \cdot x_{5,B} = 0$$

$$F_7 \cdot x_{3,C} + F_8 \cdot x_{4,C} - F_{11} \cdot x_{5,C} - F_{12} \cdot x_{5,C} = 0$$

$$F_7 \cdot x_{3,D} + F_8 \cdot x_{4,D} - F_{11} \cdot x_{5,D} - F_{12} \cdot x_{5,D} = 0$$

$$F_{23} - F_{11} - F_{12} = 0$$

$$F_5 + F_{11} + F_2 - F_9 = 0$$

$$F_5 \cdot x_{4,A} + F_{11} \cdot x_{5,A} + F_2 \cdot l_A - F_9 \cdot x_{1,A} = 0$$

$$F_5 \cdot x_{4,B} + F_{11} \cdot x_{5,B} + F_2 \cdot l_B - F_9 \cdot x_{1,B} = 0$$

$$F_5 \cdot x_{4,C} + F_{11} \cdot x_{5,C} + F_2 \cdot l_C - F_9 \cdot x_{1,C} = 0$$

$$F_5 \cdot x_{4,D} + F_{11} \cdot x_{5,D} + F_2 \cdot l_D - F_9 \cdot x_{1,D} = 0$$

$$F_9 \cdot x_{1,A} - F_9 \cdot x_{3,A} + y1 \cdot (\nu_{1,A} \cdot a_1 + \nu_{2,A} \cdot b_1 + \nu_{3,A} \cdot c_1) = 0$$

$$F_9 \cdot x_{1,B} - F_9 \cdot x_{3,B} + y1 \cdot (\nu_{1,B} \cdot a_1 + \nu_{2,B} \cdot b_1 + \nu_{3,B} \cdot c_1) = 0$$

$$F_9 \cdot x_{1,C} - F_9 \cdot x_{3,C} + y1 \cdot (\nu_{1,C} \cdot a_1 + \nu_{2,C} \cdot b_1 + \nu_{3,C} \cdot c_1) = 0$$

$$F_9 \cdot x_{1,D} - F_9 \cdot x_{3,D} + y1 \cdot (\nu_{1,D} \cdot a_1 + \nu_{2,D} \cdot b_1 + \nu_{3,D} \cdot c_1) = 0$$

$$a_1 - k_1 \cdot x_{3,A} = 0$$

$$b_1 - k_2 \cdot x_{3,B} = 0$$

$$c_1 - k_3 \cdot x_{3,A}^2 = 0$$

$$F_{13} - F_{21} + F_{16} + F_{18} + F_{20} - F_{14} - F_{22} = 0$$

$$F_{14} + F_{22} - F_{10} = 0$$

$$F_{13} \cdot x_{2,A} + F_{21} \cdot x_{7,A} + F_{16} \cdot x_{3,A} + F_{18} \cdot l_A$$
$$+ F_{20} \cdot x_{5,A} - F_{14} \cdot x_{8,A} - F_{22} \cdot x_{8,A} = 0$$

$$F_{13} \cdot x_{2,B} + F_{21} \cdot x_{7,B} + F_{16} \cdot x_{3,B} + F_{18} \cdot l_B$$
$$+ F_{20} \cdot x_{5,B} - F_{14} \cdot x_{8,B} - F_{22} \cdot x_{8,B} = 0$$

$$F_{13} \cdot x_{2,C} + F_{21} \cdot x_{7,C} + F_{16} \cdot x_{3,C} + F_{18} \cdot l_C$$
$$+ F_{20} \cdot x_{5,C} - F_{14} \cdot x_{8,C} - F_{22} \cdot x_{8,C} = 0$$

$$F_{13} \cdot x_{2,D} + F_{21} \cdot x_{7,D} + F_{16} \cdot x_{3,D} + F_{18} \cdot l_D$$
$$+ F_{20} \cdot x_{5,D} - F_{14} \cdot x_{8,D} - F_{22} \cdot x_{8,D} = 0$$

$$F_{14} \cdot x_{8,A} + F_{22} \cdot x_{4,A} - F_{10} \cdot x_{4,A} = 0$$

$$F_{14} \cdot x_{8,B} + F_{22} \cdot x_{4,B} - F_{10} \cdot x_{4,B} = 0$$

$$F_{14} \cdot x_{8,C} + F_{22} \cdot x_{4,C} - F_{10} \cdot x_{4,C} = 0$$

$$F_{14} \cdot x_{8,D} + F_{22} \cdot x_{4,D} - F_{10} \cdot x_{4,D} = 0$$

$$F_{15} + F_{17} + F_{19} + F_6 - F_{13} - F_{21} = 0$$

$$F_{15} \cdot x_{3,A} + F_{17} \cdot l_A + F_6 \cdot x_{4,A} + F_{19} \cdot x_{5,A} - F_{13} \cdot x_{2,A} - F_{21} \cdot x_{2,A} = 0$$

$$F_{15} \cdot x_{3,B} + F_{17} \cdot l_B + F_6 \cdot x_{4,B} + F_{19} \cdot x_{5,B} - F_{13} \cdot x_{2,B} - F_{21} \cdot x_{2,B} = 0$$

$$F_{15} \cdot x_{3,C} + F_{17} \cdot l_C + F_6 \cdot x_{4,C} + F_{19} \cdot x_{5,C} - F_{13} \cdot x_{2,C} - F_{21} \cdot x_{2,C} = 0$$

$$F_{15} \cdot x_{3,D} + F_{17} \cdot l_D + F_6 \cdot x_{4,D} + F_{19} \cdot x_{5,D} - F_{13} \cdot x_{2,D} - F_{21} \cdot x_{2,D} = 0$$

$$F_3 - F_{17} - F_{18} = 0$$

$$F_4 - F_{15} - F_{16} = 0$$

$$F_{12} - F_{19} - F_{20} = 0$$

$$y2 - 4 \cdot y3 = 0$$

$$F_{21} \cdot x_{2,A} - F_{21} \cdot x_{6,A} + y3 \cdot (\nu_{1,A} \cdot a_2 + \nu_{2,A} \cdot b_2 + \nu_{3,A} \cdot c_2) = 0$$

$$F_{21} \cdot x_{2,B} - F_{21} \cdot x_{6,B} + y3 \cdot (\nu_{1,B} \cdot a_2 + \nu_{2,B} \cdot b_2 + \nu_{3,B} \cdot c_2) = 0$$

$$F_{21} \cdot x_{2,C} - F_{21} \cdot x_{6,C} + y3 \cdot (\nu_{1,C} \cdot a_2 + \nu_{2,C} \cdot b_2 + \nu_{3,C} \cdot c_2) = 0$$

$$F_{21} \cdot x_{2,D} - F_{21} \cdot x_{6,D} + y3 \cdot (\nu_{1,D} \cdot a_2 + \nu_{2,D} \cdot b_2 + \nu_{3,D} \cdot c_2) = 0$$

$$F_{21} \cdot x_{6,A} - F_{21} \cdot x_{7,A} + y3 \cdot (\nu_{1,A} \cdot a_3 + \nu_{2,A} \cdot b_3 + \nu_{3,A} \cdot c_3) = 0$$

$$F_{21} \cdot x_{6,B} - F_{21} \cdot x_{7,B} + y3 \cdot (\nu_{1,B} \cdot a_3 + \nu_{2,B} \cdot b_3 + \nu_{3,B} \cdot c_3) = 0$$

$$F_{21} \cdot x_{6,C} - F_{21} \cdot x_{7,C} + y3 \cdot (\nu_{1,C} \cdot a_3 + \nu_{2,C} \cdot b_3 + \nu_{3,C} \cdot c_3) = 0$$

$$F_{21} \cdot x_{6,D} - F_{21} \cdot x_{7,D} + y3 \cdot (\nu_{1,D} \cdot a_3 + \nu_{2,D} \cdot b_3 + \nu_{3,D} \cdot c_3) = 0$$

$$a_2 - k_1 \cdot x_{6,A} = 0$$

$$b_2 - k_2 \cdot x_{6,B} = 0$$

$$c_2 - k_3 \cdot x_{6,A}^2 = 0$$

$$a_3 - k_1 \cdot x_{7,A} = 0$$

$$b_3 - k_2 \cdot x_{7,B} = 0$$

$$c_3 - k_3 \cdot x_{7,A}^2 = 0$$

$$F_{22} \cdot x_{8,A} - F_{22} \cdot x_{9,A} + y3 \cdot (\nu_{1,A} \cdot a_4 + \nu_{2,A} \cdot b_4 + \nu_{3,A} \cdot c_4) = 0$$

$$F_{22} \cdot x_{8,B} - F_{22} \cdot x_{9,B} + y3 \cdot (\nu_{1,B} \cdot a_4 + \nu_{2,B} \cdot b_4 + \nu_{3,B} \cdot c_4) = 0$$

$$F_{22} \cdot x_{8,C} - F_{22} \cdot x_{9,C} + y3 \cdot (\nu_{1,C} \cdot a_4 + \nu_{2,C} \cdot b_4 + \nu_{3,C} \cdot c_4) = 0$$

$$F_{22} \cdot x_{8,D} - F_{22} \cdot x_{9,D} + y3 \cdot (\nu_{1,D} \cdot a_4 + \nu_{2,D} \cdot b_4 + \nu_{3,D} \cdot c_4) = 0$$

$$F_{22} \cdot x_{9,A} - F_{22} \cdot x_{4,A} + y3 \cdot (\nu_{1,A} \cdot a_5 + \nu_{2,A} \cdot b_5 + \nu_{3,A} \cdot c_5) = 0$$

$$F_{22} \cdot x_{9,B} - F_{22} \cdot x_{4,B} + y3 \cdot (\nu_{1,B} \cdot a_5 + \nu_{2,B} \cdot b_5 + \nu_{3,B} \cdot c_5) = 0$$

$$F_{22} \cdot x_{9,C} - F_{22} \cdot x_{4,C} + y3 \cdot (\nu_{1,C} \cdot a_5 + \nu_{2,C} \cdot b_5 + \nu_{3,C} \cdot c_5) = 0$$

$$F_{22} \cdot x_{9,D} - F_{22} \cdot x_{4,D} + y3 \cdot (\nu_{1,D} \cdot a_5 + \nu_{2,D} \cdot b_5 + \nu_{3,D} \cdot c_5) = 0$$

$$a_4 - k_1 \cdot x_{9,A} = 0$$

$$b_4 - k_2 \cdot x_{9,B} = 0$$

$$c_4 - k_3 \cdot x_{9,A}^2 = 0$$

$$a_5 - k_1 \cdot x_{4,A} = 0$$

$$b_5 - k_2 \cdot x_{4,B} = 0$$

$$c_5 - k_3 \cdot x_{4,A}^2 = 0$$

9.2.2 Data

- inlet concentrations: l_m

	l_m
A	5.8
B	0
C	0
D	0

- stoichiometric coefficients: $\nu_{rp,m}$

	A	B	C	D
1	-1	1	0	0
2	0	-1	1	0
3	-1	0	0	1

- kinetic constants: k_i

	k_i
1	10
2	1
3	1

• prespecified streams

	Value
F_1	100
F_{12}	100

• Upper and lower bounds

Variable		Lower	Upper
F_i	$i = 2, 3, 17, 18$	0.0	100
F_i	$i = 4, \ldots, 16, 19, \ldots, 22$	0.0	5,000
$x_{i,m}$	$m = A, B, C, D \quad i = 1, \ldots, 9$	0.0	5.8
$y1$		0.1	250
$y3$		0.001	4.0

9.2.3 Problem Statistics

Number of Variables	76
Number of Linear Constraints	23
Number of Nonlinear Constraints	45

9.2.4 Best Known Solution

• Objective value=3.5793

• Flowrates

$$F_i = 100 \qquad i = 1, 2, 4, 8, 9, 10, 15, 21, 22$$

- Concentrations

Var.	Level	Var.	Level	Var.	Level	Var.	Level
$x_{1,A}$	5.80	$x_{2,A}$	2.157	$x_{3,A}$	2.157	$x_{4,A}$	0.506
$x_{1,B}$	0.0	$x_{2,B}$	2.631	$x_{3,B}$	2.631	$x_{4,B}$	3.579
$x_{1,C}$	0.0	$x_{2,C}$	0.365	$x_{3,C}$	0.365	$x_{4,C}$	0.908
$x_{1,D}$	0.0	$x_{2,D}$	0.646	$x_{3,D}$	0.646	$x_{4,D}$	0.807
$x_{5,A}$	0.506	$x_{6,A}$	1.479	$x_{7,A}$	1.026	$x_{8,A}$	1.026
$x_{5,B}$	3.579	$x_{6,B}$	3.098	$x_{7,B}$	3.374	$x_{8,B}$	3.374
$x_{5,C}$	0.908	$x_{6,C}$	0.489	$x_{7,C}$	0.624	$x_{8,C}$	0.624
$x_{5,D}$	0.807	$x_{6,D}$	0.734	$x_{7,D}$	0.776	$x_{8,D}$	0.776
$x_{9,A}$	0.718	$x_{9,B}$	3.520	$x_{9,C}$	0.765	$x_{9,D}$	0.797

- Reaction rates and volumes

Var.	Level	Var.	Level	Var.	Level
a_1	21.575	b_1	2.631	c_1	4.655
a_2	14.786	b_2	3.098	c_2	2.186
a_3	10.261	b_3	3.374	c_3	1.053
a_4	7.182	b_4	3.520	c_4	0.516
a_5	5.057	b_5	3.579	c_5	0.256
$y1$	13.887	$y2$	16.0	$y3$	4.0

9.3 Test Problem 2 : Trambouze and Piret reaction

This test problem is taken from [125] and features the Trambouze reaction :

- *reaction mechanism:*

$$A \xrightarrow{k_1} B \quad (zero\ order\ reaction) \tag{9.4}$$

$$A \xrightarrow{k_2} C \quad (first\ order\ reaction) \tag{9.5}$$

$$A \xrightarrow{k_3} D \quad (second\ order\ reaction) \tag{9.6}$$

- *objective function:* Selectivity of C over A

9.3.1 Problem Formulation

$$\text{MAX} \quad \frac{x_{5,C}}{l_A - x_{5,A}}$$

$$F_1 - F_2 - F_3 = 0$$

$$F_4 + F_7 - F_9 = 0$$

$$F_5 + F_6 + F_8 - F_{10} = 0$$

$$F_7 + F_8 - F_{23} - F_{12} = 0$$

$$F_7 \cdot x_{3,A} + F_8 \cdot x_{4,A} - F_{11} \cdot x_{5,A} - F_{12} \cdot x_{5,A} = 0$$

$$F_7 \cdot x_{3,B} + F_8 \cdot x_{4,B} - F_{11} \cdot x_{5,B} - F_{12} \cdot x_{5,B} = 0$$

$$F_7 \cdot x_{3,C} + F_8 \cdot x_{4,C} - F_{11} \cdot x_{5,C} - F_{12} \cdot x_{5,C} = 0$$

$$F_7 \cdot x_{3,D} + F_8 \cdot x_{4,D} - F_{11} \cdot x_{5,D} - F_{12} \cdot x_{5,D} = 0$$

$$F_{23} - F_{11} - F_{12} = 0$$

$$F_5 + F_{11} + F_2 - F_9 = 0$$

$$F_5 \cdot x_{4,A} + F_{11} \cdot x_{5,A} + F_2 \cdot l_A - F_9 \cdot x_{1,A} = 0$$

$$F_5 \cdot x_{4,B} + F_{11} \cdot x_{5,B} + F_2 \cdot l_B - F_9 \cdot x_{1,B} = 0$$

$$F_5 \cdot x_{4,C} + F_{11} \cdot x_{5,C} + F_2 \cdot l_C - F_9 \cdot x_{1,C} = 0$$

$$F_5 \cdot x_{4,D} + F_{11} \cdot x_{5,D} + F_2 \cdot l_D - F_9 \cdot x_{1,D} = 0$$

$$F_9 \cdot x_{1,A} - F_9 \cdot x_{3,A} + y1 \cdot (\nu_{1,A} \cdot k_1 + \nu_{2,A} \cdot b_1 + \nu_{3,A} \cdot c_1) = 0$$

$$F_9 \cdot x_{1,B} - F_9 \cdot x_{3,B} + y1 \cdot (\nu_{1,B} \cdot k_1 + \nu_{2,B} \cdot b_1 + \nu_{3,B} \cdot c_1) = 0$$

$$F_9 \cdot x_{1,C} - F_9 \cdot x_{3,C} + y1 \cdot (\nu_{1,C} \cdot k_1 + \nu_{2,C} \cdot b_1 + \nu_{3,C} \cdot c_1) = 0$$

$$F_9 \cdot x_{1,D} - F_9 \cdot x_{3,D} + y1 \cdot (\nu_{1,D} \cdot k_1 + \nu_{2,D} \cdot b_1 + \nu_{3,D} \cdot c_1) = 0$$

$$b_1 - k_2 \cdot x_{3,A} = 0$$

$$c_1 - k_3 \cdot x_{3,A}^2 = 0$$

$$F_{13} - F_{21} + F_{16} + F_{18} + F_{20} - F_{14} - F_{22} = 0$$

$$F_{14} + F_{22} - F_{10} = 0$$

$$F_{13} \cdot x_{2,A} + F_{21} \cdot x_{7,A} + F_{16} \cdot x_{3,A} + F_{18} \cdot l_A$$

$$+ F_{20} \cdot x_{5,A} - F_{14} \cdot x_{8,A} - F_{22} \cdot x_{8,A} = 0$$

$$F_{13} \cdot x_{2,B} + F_{21} \cdot x_{7,B} + F_{16} \cdot x_{3,B} + F_{18} \cdot l_B$$
$$+F_{20} \cdot x_{5,B} - F_{14} \cdot x_{8,B} - F_{22} \cdot x_{8,B} = 0$$
$$F_{13} \cdot x_{2,C} + F_{21} \cdot x_{7,C} + F_{16} \cdot x_{3,C} + F_{18} \cdot l_C$$
$$+F_{20} \cdot x_{5,C} - F_{14} \cdot x_{8,C} - F_{22} \cdot x_{8,C} = 0$$
$$F_{13} \cdot x_{2,D} + F_{21} \cdot x_{7,D} + F_{16} \cdot x_{3,D} + F_{18} \cdot l_D$$
$$+F_{20} \cdot x_{5,D} - F_{14} \cdot x_{8,D} - F_{22} \cdot x_{8,D} = 0$$
$$F_{14} \cdot x_{8,A} + F_{22} \cdot x_{4,A} - F_{10} \cdot x_{4,A} = 0$$
$$F_{14} \cdot x_{8,B} + F_{22} \cdot x_{4,B} - F_{10} \cdot x_{4,B} = 0$$
$$F_{14} \cdot x_{8,C} + F_{22} \cdot x_{4,C} - F_{10} \cdot x_{4,C} = 0$$
$$F_{14} \cdot x_{8,D} + F_{22} \cdot x_{4,D} - F_{10} \cdot x_{4,D} = 0$$
$$F_{15} + F_{17} + F_6 + F_{19} - F_{13} - F_{21} = 0$$
$$F_{15} \cdot x_{3,A} + F_{17} \cdot l_A + F_6 \cdot x_{4,A} + F_{19} \cdot x_{5,A} - F_{13} \cdot x_{2,A} - F_{21} \cdot x_{2,A} = 0$$
$$F_{15} \cdot x_{3,B} + F_{17} \cdot l_B + F_6 \cdot x_{4,B} + F_{19} \cdot x_{5,B} - F_{13} \cdot x_{2,B} - F_{21} \cdot x_{2,B} = 0$$
$$F_{15} \cdot x_{3,C} + F_{17} \cdot l_C + F_6 \cdot x_{4,C} + F_{19} \cdot x_{5,C} - F_{13} \cdot x_{2,C} - F_{21} \cdot x_{2,C} = 0$$
$$F_{15} \cdot x_{3,D} + F_{17} \cdot l_D + F_6 \cdot x_{4,D} + F_{19} \cdot x_{5,D} - F_{13} \cdot x_{2,D} - F_{21} \cdot x_{2,D} = 0$$
$$F_3 - F_{17} - F_{18} = 0$$
$$F_4 - F_{15} - F_{16} = 0$$
$$F_{12} - F_{19} - F_{20} = 0$$
$$y2 - 4 \cdot y3 = 0$$
$$F_{21} \cdot x_{2,A} - F_{21} \cdot x_{6,A} + y3 \cdot (\nu_{1,A} \cdot k_1 + \nu_{2,A} \cdot b_2 + \nu_{3,A} \cdot c_2) = 0$$
$$F_{21} \cdot x_{2,B} - F_{21} \cdot x_{6,B} + y3 \cdot (\nu_{1,B} \cdot k_1 + \nu_{2,B} \cdot b_2 + \nu_{3,B} \cdot c_2) = 0$$
$$F_{21} \cdot x_{2,C} - F_{21} \cdot x_{6,C} + y3 \cdot (\nu_{1,C} \cdot k_1 + \nu_{2,C} \cdot b_2 + \nu_{3,C} \cdot c_2) = 0$$
$$F_{21} \cdot x_{2,D} - F_{21} \cdot x_{6,D} + y3 \cdot (\nu_{1,D} \cdot k_1 + \nu_{2,D} \cdot b_2 + \nu_{3,D} \cdot c_2) = 0$$
$$F_{21} \cdot x_{6,A} - F_{21} \cdot x_{7,A} + y3 \cdot (\nu_{1,A} \cdot k_1 + \nu_{2,A} \cdot b_3 + \nu_{3,A} \cdot c_3) = 0$$
$$F_{21} \cdot x_{6,B} - F_{21} \cdot x_{7,B} + y3 \cdot (\nu_{1,B} \cdot k_1 + \nu_{2,B} \cdot b_3 + \nu_{3,B} \cdot c_3) = 0$$
$$F_{21} \cdot x_{6,C} - F_{21} \cdot x_{7,C} + y3 \cdot (\nu_{1,C} \cdot k_1 + \nu_{2,C} \cdot b_3 + \nu_{3,C} \cdot c_3) = 0$$
$$F_{21} \cdot x_{6,D} - F_{21} \cdot x_{7,D} + y3 \cdot (\nu_{1,D} \cdot k_1 + \nu_{2,D} \cdot b_3 + \nu_{3,D} \cdot c_3) = 0$$

$$b_2 - k_2 \cdot x_{6,A} = 0$$

$$c_2 - k_3 \cdot x_{6,A}^2 = 0$$

$$b_3 - k_2 \cdot x_{7,A} = 0$$

$$c_3 - k_3 \cdot x_{7,A}^2 = 0$$

$$F_{22} \cdot x_{8,A} - F_{22} \cdot x_{9,A} + y3 \cdot (\nu_{1,A} \cdot k_1 + \nu_{2,A} \cdot b_4 + \nu_{3,A} \cdot c_4) = 0$$

$$F_{22} \cdot x_{8,B} - F_{22} \cdot x_{9,B} + y3 \cdot (\nu_{1,B} \cdot k_1 + \nu_{2,B} \cdot b_4 + \nu_{3,B} \cdot c_4) = 0$$

$$F_{22} \cdot x_{8,C} - F_{22} \cdot x_{9,C} + y3 \cdot (\nu_{1,C} \cdot k_1 + \nu_{2,C} \cdot b_4 + \nu_{3,C} \cdot c_4) = 0$$

$$F_{22} \cdot x_{8,D} - F_{22} \cdot x_{9,D} + y3 \cdot (\nu_{1,D} \cdot k_1 + \nu_{2,D} \cdot b_4 + \nu_{3,D} \cdot c_4) = 0$$

$$F_{22} \cdot x_{9,A} - F_{22} \cdot x_{4,A} + y3 \cdot (\nu_{1,A} \cdot k_1 + \nu_{2,A} \cdot b_5 + \nu_{3,A} \cdot c_5) = 0$$

$$F_{22} \cdot x_{9,B} - F_{22} \cdot x_{4,B} + y3 \cdot (\nu_{1,B} \cdot k_1 + \nu_{2,B} \cdot b_5 + \nu_{3,B} \cdot c_5) = 0$$

$$F_{22} \cdot x_{9,C} - F_{22} \cdot x_{4,C} + y3 \cdot (\nu_{1,C} \cdot k_1 + \nu_{2,C} \cdot b_5 + \nu_{3,C} \cdot c_5) = 0$$

$$F_{22} \cdot x_{9,D} - F_{22} \cdot x_{4,D} + y3 \cdot (\nu_{1,D} \cdot k_1 + \nu_{2,D} \cdot b_5 + \nu_{3,D} \cdot c_5) = 0$$

$$b_4 - k_2 \cdot x_{9,A} = 0$$

$$c_4 - k_3 \cdot x_{9,A}^2 = 0$$

$$b_5 - k_2 \cdot x_{4,A} = 0$$

$$c_5 - k_3 \cdot x_{4,A}^2 = 0$$

9.3.2 Data

- inlet compositions: l_m

	l_m
A	1.0
B	0
C	0
D	0

- stoichiometric coefficients: $\nu_{rp,m}$

	A	B	C	D
1	-1	1	0	0
2	-1	0	1	0
3	-1	0	0	1

- kinetic constants: k_i

	k_i
1	0.025
2	0.2
3	0.4

- prespecified streams

	Value
F_1	100
F_{12}	100

- Upper and lower bounds

Variable			Lower	Upper
F_i		$i = 2, 3, 17, 18$	0.0	100
F_i		$i = 4, \ldots, 16, 19, \ldots, 22$	0.0	5,000
$x_{i,m}$	$m = A, B, C, D$	$i = 1, \ldots, 9$	0.0	2.0
$y1$			1	5,000
$y3$			0.001	100.0

9.3.3 Problem Statistics

Number of Variables	71
Number of Linear Constraints	18
Number of Nonlinear Constraints	45

9.3.4 Best Known Solution

- Objective value=0.500

- Flowrates

Var.	Level	Var.	Level	Var.	Level	Var.	Level
F_1	100.0	F_2	98.593	F_3	1.407	F_4	98.742
F_5	0.149	F_6	18.455	F_7	0.0	F_8	100.0
F_9	98.742	F_{10}	93.461	F_{11}	0.0	F_{12}	0.0
F_{13}	0.0	F_{14}	0.0	F_{15}	74.399	F_{16}	24.365
F_{17}	0.629	F_{18}	0.778	F_{19}	0.0	F_{20}	0.0
F_{21}	93.461	F_{22}	118.604				

- Concentrations

Var.	Level	Var.	Level	Var.	Level	Var.	Level
$x_{1,A}$	0.999	$x_{2,A}$	0.255	$x_{3,A}$	0.250	$x_{4,A}$	0.249
$x_{1,B}$	2.82 E-4	$x_{2,B}$	0.186	$x_{3,B}$	0.187	$x_{4,B}$	0.188
$x_{1,C}$	5.65 E-4	$x_{2,C}$	0.373	$x_{3,C}$	0.375	$x_{4,C}$	0.376
$x_{1,D}$	2.82 E-4	$x_{2,D}$	0.186	$x_{3,D}$	0.188	$x_{4,D}$	0.188
$x_{5,A}$	0.249	$x_{6,A}$	0.252	$x_{7,A}$	0.254	$x_{8,A}$	0.251
$x_{5,B}$	0.188	$x_{6,B}$	0.187	$x_{7,B}$	0.187	$x_{8,B}$	0.187
$x_{5,C}$	0.376	$x_{6,C}$	0.374	$x_{7,C}$	0.373	$x_{8,C}$	0.374
$x_{5,D}$	0.188	$x_{6,D}$	0.187	$x_{7,D}$	0.187	$x_{8,D}$	0.187
$x_{9,A}$	0.249	$x_{9,B}$	0.188	$x_{9,C}$	0.376	$x_{9,D}$	0.188

- Reaction rates and volumes

Var.	Level	Var.	Level	Var.	Level
b_1	0.050	b_2	0.050	b_3	0.050
b_4	0.050	b_5	0.050	c_1	0.025
c_2	0.025	c_3	0.025	c_4	0.025
c_5	0.025	$y1$	739.405	$y2$	11.660
$y3$	2.915				

9.4 Test Problem 3 : Fuguitt and Hawkins reaction

This test problem is taken from [125] and involves the Fuguitt and Hawkins reaction :

- *reaction mechanism:*

$$A \xrightarrow{k_1} E \quad (first \ order \ reaction \ ) \tag{9.7}$$

$$A \xrightarrow{k_2} D \quad (first \ order \ reaction \ ) \tag{9.8}$$

$$2 A \xrightarrow{k_5} C \quad (second \ order \ reaction \ ) \tag{9.9}$$

$$C \xrightleftharpoons{k_4} 2 D \quad (first/second \ order \ reaction \ ) \tag{9.10}$$

$$B \xrightleftharpoons{k_3} D \quad (first \ order \ reaction \ ) \tag{9.11}$$

$$\tag{9.12}$$

- *objective function:* Selectivity of C over D.

9.4.1   Problem Formulation

$$\text{MAX} \quad \frac{x_{5,C}}{x_{5,D}}$$

$$F_1 - F_2 - F_3 = 0$$

$$F_4 + F_7 - F_9 = 0$$

$$F_5 + F_6 + F_8 - F_{10} = 0$$

$$F_7 + F_8 - F_{23} - F_{12} = 0$$

$$F_7 \cdot x_{3,A} + F_8 \cdot x_{4,A} - F_{11} \cdot x_{5,A} - F_{12} \cdot x_{5,A} = 0$$

$$F_7 \cdot x_{3,B} + F_8 \cdot x_{4,B} - F_{11} \cdot x_{5,B} - F_{12} \cdot x_{5,B} = 0$$

$$F_7 \cdot x_{3,C} + F_8 \cdot x_{4,C} - F_{11} \cdot x_{5,C} - F_{12} \cdot x_{5,C} = 0$$

$$F_7 \cdot x_{3,D} + F_8 \cdot x_{4,D} - F_{11} \cdot x_{5,D} - F_{12} \cdot x_{5,D} = 0$$

$$F_7 \cdot x_{3,E} + F_8 \cdot x_{4,E} - F_{11} \cdot x_{5,E} - F_{12} \cdot x_{5,E} = 0$$

$$F_{23} - F_{11} - F_{12} = 0$$

$$F_5 + F_{11} + F_2 - F_9 = 0$$

$$F_5 \cdot x_{4,A} + F_{11} \cdot x_{5,A} + F_2 \cdot l_A - F_9 \cdot x_{1,A} = 0$$

$$F_5 \cdot x_{4,B} + F_{11} \cdot x_{5,B} + F_2 \cdot l_B - F_9 \cdot x_{1,B} = 0$$

$$F_5 \cdot x_{4,C} + F_{11} \cdot x_{5,C} + F_2 \cdot l_C - F_9 \cdot x_{1,C} = 0$$

$$F_5 \cdot x_{4,D} + F_{11} \cdot x_{5,D} + F_2 \cdot l_D - F_9 \cdot x_{1,D} = 0$$

$$F_5 \cdot x_{4,E} + F_{11} \cdot x_{5,E} + F_2 \cdot l_E - F_9 \cdot x_{1,E} = 0$$

$$F_9 \cdot x_{1,A} - F_9 \cdot x_{3,A} + y1 \cdot (\nu_{1,A} \cdot a_1 +$$

$$\nu_{2,A} \cdot b_1 + \nu_{3,A} \cdot c_1 + \nu_{4,A} \cdot d_1 + \nu_{5,A} \cdot e_1 + \nu_{6,A} \cdot f_1 + \nu_{7,A} \cdot g_1) = 0$$

$$F_9 \cdot x_{1,B} - F_9 \cdot x_{3,B} + y1 \cdot (\nu_{1,B} \cdot a_1 +$$

$$\nu_{2,B} \cdot b_1 + \nu_{3,B} \cdot c_1 + \nu_{4,B} \cdot d_1 + \nu_{5,B} \cdot e_1 + \nu_{6,B} \cdot f_1 + \nu_{7,B} \cdot g_1) = 0$$

$$F_9 \cdot x_{1,C} - F_9 \cdot x_{3,C} + y1 \cdot (\nu_{1,C} \cdot a_1$$

$$\nu_{2,C} \cdot b_1 + \nu_{3,C} \cdot c_1 + \nu_{4,C} \cdot d_1 + \nu_{5,C} \cdot e_1 + \nu_{6,C} \cdot f_1 + \nu_{7,C} \cdot g_1) = 0$$

$$F_9 \cdot x_{1,D} - F_9 \cdot x_{3,D} + y1 \cdot (\nu_{1,D} \cdot a_1 +$$

$$\nu_{2,D} \cdot b_1 + \nu_{3,D} \cdot c_1 + \nu_{4,D} \cdot d_1 + \nu_{5,D} \cdot e_1 + \nu_{6,D} \cdot f_1 + \nu_{7,D} \cdot g_1) = 0$$

$$F_9 \cdot x_{1,E} - F_9 \cdot x_{3,E} + y1 \cdot (\nu_{1,E} \cdot a_1$$

$$\nu_{2,E} \cdot b_1 + \nu_{3,E} \cdot c_1 + \nu_{4,E} \cdot d_1 + \nu_{5,E} \cdot e_1 + \nu_{6,E} \cdot f_1 + \nu_{7,E} \cdot g_1) = 0$$

$$a_1 - k_1 \cdot x_{3,A} = 0$$

$$b_1 - k_2 \cdot x_{3,A} = 0$$

$$c_1 - k_3 \cdot x_{3,D} = 0$$

$$d_1 - k_4 \cdot x_{3,D}^2 = 0$$

$$e_1 - k_5 \cdot x_{3,A}^2 = 0$$

$$f_1 - k_6 \cdot x_{3,B} = 0$$

$$g_1 - k_7 \cdot x_{3,C} = 0$$

$$F_{13} - F_{21} + F_{16} + F_{18} + F_{20} - F_{14} - F_{22} = 0$$

$$F_{14} + F_{22} - F_{10} = 0$$

$$F_{13} \cdot x_{2,A} + F_{21} \cdot x_{7,A} + F_{16} \cdot x_{3,A} + F_{18} \cdot l_A$$

$$+ F_{20} \cdot x_{5,A} - F_{14} \cdot x_{8,A} - F_{22} \cdot x_{8,A} = 0$$

$$F_{13} \cdot x_{2,B} + F_{21} \cdot x_{7,B} + F_{16} \cdot x_{3,B} + F_{18} \cdot l_B$$

$$+ F_{20} \cdot x_{5,B} - F_{14} \cdot x_{8,B} - F_{22} \cdot x_{8,B} = 0$$

$$F_{13} \cdot x_{2,C} + F_{21} \cdot x_{7,C} + F_{16} \cdot x_{3,C} + F_{18} \cdot l_C$$

$$+ F_{20} \cdot x_{5,C} - F_{14} \cdot x_{8,C} - F_{22} \cdot x_{8,C} = 0$$

$$F_{13} \cdot x_{2,D} + F_{21} \cdot x_{7,D} + F_{16} \cdot x_{3,D} + F_{18} \cdot l_D$$
$$+ F_{20} \cdot x_{5,D} - F_{14} \cdot x_{8,D} - F_{22} \cdot x_{8,D} = 0$$
$$F_{13} \cdot x_{2,E} + F_{21} \cdot x_{7,E} + F_{16} \cdot x_{3,E} + F_{18} \cdot l_E$$
$$+ F_{20} \cdot x_{5,E} - F_{14} \cdot x_{8,E} - F_{22} \cdot x_{8,E} = 0$$
$$F_{14} \cdot x_{8,A} + F_{22} \cdot x_{4,A} - F_{10} \cdot x_{4,A} = 0$$
$$F_{14} \cdot x_{8,B} + F_{22} \cdot x_{4,B} - F_{10} \cdot x_{4,B} = 0$$
$$F_{14} \cdot x_{8,C} + F_{22} \cdot x_{4,C} - F_{10} \cdot x_{4,C} = 0$$
$$F_{14} \cdot x_{8,D} + F_{22} \cdot x_{4,D} - F_{10} \cdot x_{4,D} = 0$$
$$F_{14} \cdot x_{8,E} + F_{22} \cdot x_{4,E} - F_{10} \cdot x_{4,E} = 0$$
$$F_{15} + F_{17} + F_{19} + F_6 - F_{13} - F_{21} = 0$$
$$F_{15} \cdot x_{3,A} + F_{17} \cdot l_A + F_6 \cdot x_{4,A} + F_{19} \cdot x_{5,A} - F_{13} \cdot x_{2,A} - F_{21} \cdot x_{2,A} = 0$$
$$F_{15} \cdot x_{3,B} + F_{17} \cdot l_B + F_6 \cdot x_{4,B} + F_{19} \cdot x_{5,B} - F_{13} \cdot x_{2,B} - F_{21} \cdot x_{2,B} = 0$$
$$F_{15} \cdot x_{3,C} + F_{17} \cdot l_C + F_6 \cdot x_{4,C} + F_{19} \cdot x_{5,C} - F_{13} \cdot x_{2,C} - F_{21} \cdot x_{2,C} = 0$$
$$F_{15} \cdot x_{3,D} + F_{17} \cdot l_D + F_6 \cdot x_{4,D} + F_{19} \cdot x_{5,D} - F_{13} \cdot x_{2,D} - F_{21} \cdot x_{2,D} = 0$$
$$F_{15} \cdot x_{3,E} + F_{17} \cdot l_E + F_6 \cdot x_{4,E} + F_{19} \cdot x_{5,E} - F_{13} \cdot x_{2,E} - F_{21} \cdot x_{2,E} = 0$$
$$F_3 - F_{17} - F_{18} = 0$$
$$F_4 - F_{15} - F_{16} = 0$$
$$F_{12} - F_{19} - F_{20} = 0$$
$$y2 - 4 \cdot y3 = 0$$
$$F_{21} \cdot x_{2,A} - F_{21} \cdot x_{6,A} + y3 \cdot (\nu_{1,A} \cdot a_2$$
$$+ \nu_{2,A} \cdot b_2 + \nu_{3,A} \cdot c_2 + \nu_{4,A} \cdot d_2 + \nu_{5,A} \cdot e_2 + \nu_{6,A} \cdot f_2 + \nu_{7,A} \cdot g_2) = 0$$
$$F_{21} \cdot x_{2,B} - F_{21} \cdot x_{6,B} + y3 \cdot (\nu_{1,B} \cdot a_2$$
$$+ \nu_{2,B} \cdot b_2 + \nu_{3,B} \cdot c_2 + \nu_{4,B} \cdot d_2 + \nu_{5,B} \cdot e_2 + \nu_{6,B} \cdot f_2 + \nu_{7,B} \cdot g_2) = 0$$
$$F_{21} \cdot x_{2,C} - F_{21} \cdot x_{6,C} + y3 \cdot (\nu_{1,C} \cdot a_2$$
$$+ \nu_{2,C} \cdot b_2 + \nu_{3,C} \cdot c_2 + \nu_{4,C} \cdot d_2 + \nu_{5,C} \cdot e_2 + \nu_{6,C} \cdot f_2 + \nu_{7,C} \cdot g_2) = 0$$
$$F_{21} \cdot x_{2,D} - F_{21} \cdot x_{6,D} + y3 \cdot (\nu_{1,D} \cdot a_2$$
$$+ \nu_{2,D} \cdot b_2 + \nu_{3,D} \cdot c_2 + \nu_{4,D} \cdot d_2 + \nu_{5,D} \cdot e_2 + \nu_{6,D} \cdot f_2 + \nu_{7,D} \cdot g_2) = 0$$

$$F_{21} \cdot x_{2,E} - F_{21} \cdot x_{6,E} + y3 \cdot (\nu_{1,E} \cdot a_2$$
$$+\nu_{2,E} \cdot b_2 + \nu_{3,E} \cdot c_2 + \nu_{4,E} \cdot d_2 + \nu_{5,E} \cdot e_2 + \nu_{6,E} \cdot f_2 + \nu_{7,E} \cdot g_2) = 0$$
$$F_{21} \cdot x_{6,A} - F_{21} \cdot x_{7,A} + y3 \cdot (\nu_{1,A} \cdot a_3$$
$$+\nu_{2,A} \cdot b_3 + \nu_{3,A} \cdot c_3 + \nu_{4,A} \cdot d_3 + \nu_{5,A} \cdot e_3 + \nu_{6,A} \cdot f_3 + \nu_{7,A} \cdot g_3) = 0$$
$$F_{21} \cdot x_{6,B} - F_{21} \cdot x_{7,B} + y3 \cdot (\nu_{1,B} \cdot a_3$$
$$+\nu_{2,B} \cdot b_3 + \nu_{3,B} \cdot c_3 + \nu_{4,B} \cdot d_3 + \nu_{5,B} \cdot e_3 + \nu_{6,B} \cdot f_3 + \nu_{7,B} \cdot g_3) = 0$$
$$F_{21} \cdot x_{6,C} - F_{21} \cdot x_{7,C} + y3 \cdot (\nu_{1,C} \cdot a_3$$
$$+\nu_{2,C} \cdot b_3 + \nu_{3,C} \cdot c_3 + \nu_{4,C} \cdot d_3 + \nu_{5,C} \cdot e_3 + \nu_{6,C} \cdot f_3 + \nu_{7,C} \cdot g_3) = 0$$
$$F_{21} \cdot x_{6,D} - F_{21} \cdot x_{7,D} + y3 \cdot (\nu_{1,D} \cdot a_3$$
$$+\nu_{2,D} \cdot b_3 + \nu_{3,D} \cdot c_3 + \nu_{4,D} \cdot d_3 + \nu_{5,D} \cdot e_3 + \nu_{6,D} \cdot f_3 + \nu_{7,D} \cdot g_3) = 0$$
$$F_{21} \cdot x_{6,E} - F_{21} \cdot x_{7,E} + y3 \cdot (\nu_{1,E} \cdot a_3$$
$$+\nu_{2,E} \cdot b_3 + \nu_{3,E} \cdot c_3 + \nu_{4,E} \cdot d_3 + \nu_{5,E} \cdot e_3 + \nu_{6,E} \cdot f_3 + \nu_{7,E} \cdot g_3) = 0$$
$$a_2 - k_1 \cdot x_{6,A} = 0$$
$$b_2 - k_2 \cdot x_{6,A} = 0$$
$$c_2 - k_3 \cdot x_{6,D} = 0$$
$$d_2 - k_4 \cdot x_{6,D}^2 = 0$$
$$e_2 - k_5 \cdot x_{6,A}^2 = 0$$
$$f_2 - k_6 \cdot x_{6,B} = 0$$
$$g_2 - k_7 \cdot x_{6,C} = 0$$
$$a_3 - k_1 \cdot x_{7,A} = 0$$
$$b_3 - k_2 \cdot x_{7,A} = 0$$
$$c_3 - k_3 \cdot x_{7,D} = 0$$
$$d_3 - k_4 \cdot x_{7,D}^2 = 0$$
$$e_3 - k_5 \cdot x_{7,A}^2 = 0$$
$$f_3 - k_6 \cdot x_{7,B} = 0$$
$$g_3 - k_7 \cdot x_{7,C} = 0$$
$$F_{22} \cdot x_{8,A} - F_{22} \cdot x_{9,A} + y3 \cdot (\nu_{1,A} \cdot a_4$$

$$+\nu_{2,A} \cdot b_4 + \nu_{3,A} \cdot c_4 + \nu_{4,A} \cdot d_4 + \nu_{5,A} \cdot e_4 + \nu_{6,A} \cdot f_4 + \nu_{7,A} \cdot g_4) = 0$$

$$F_{22} \cdot x_{8,B} - F_{22} \cdot x_{9,B} + y3 \cdot (\nu_{1,B} \cdot a_4$$

$$+\nu_{2,B} \cdot b_4 + \nu_{3,B} \cdot c_4 + \nu_{4,B} \cdot d_4 + \nu_{5,B} \cdot e_4 + \nu_{6,B} \cdot f_4 + \nu_{7,B} \cdot g_4) = 0$$

$$F_{22} \cdot x_{8,C} - F_{22} \cdot x_{9,C} + y3 \cdot (\nu_{1,C} \cdot a_4$$

$$+\nu_{2,C} \cdot b_4 + \nu_{3,C} \cdot c_4 + \nu_{4,C} \cdot d_4 + \nu_{5,C} \cdot e_4 + \nu_{6,C} \cdot f_4 + \nu_{7,C} \cdot g_4) = 0$$

$$F_{22} \cdot x_{8,D} - F_{22} \cdot x_{9,D} + y3 \cdot (\nu_{1,D} \cdot a_4$$

$$+\nu_{2,D} \cdot b_4 + \nu_{3,D} \cdot c_4 + \nu_{4,D} \cdot d_4 + \nu_{5,D} \cdot e_4 + \nu_{6,D} \cdot f_4 + \nu_{7,D} \cdot g_4) = 0$$

$$F_{22} \cdot x_{8,E} - F_{22} \cdot x_{9,E} + y3 \cdot (\nu_{1,E} \cdot a_4$$

$$+\nu_{2,E} \cdot b_4 + \nu_{3,E} \cdot c_4 + \nu_{4,E} \cdot d_4 + \nu_{5,E} \cdot e_4 + \nu_{6,E} \cdot f_4 + \nu_{7,E} \cdot g_4) = 0$$

$$F_{22} \cdot x_{9,A} - F_{22} \cdot x_{4,A} + y3 \cdot (\nu_{1,A} \cdot a_5$$

$$+\nu_{2,A} \cdot b_5 + \nu_{3,A} \cdot c_5 + \nu_{4,A} \cdot d_5 + \nu_{5,A} \cdot e_5 + \nu_{6,A} \cdot f_5 + \nu_{7,A} \cdot g_5) = 0$$

$$F_{22} \cdot x_{9,B} - F_{22} \cdot x_{4,B} + y3 \cdot (\nu_{1,B} \cdot a_5$$

$$+\nu_{2,B} \cdot b_5 + \nu_{3,B} \cdot c_5 + \nu_{4,B} \cdot d_5 + \nu_{5,B} \cdot e_5 + \nu_{6,B} \cdot f_5 + \nu_{7,B} \cdot g_5) = 0$$

$$F_{22} \cdot x_{9,C} - F_{22} \cdot x_{4,C} + y3 \cdot (\nu_{1,C} \cdot a_5$$

$$+\nu_{2,C} \cdot b_5 + \nu_{3,C} \cdot c_5 + \nu_{4,C} \cdot d_5 + \nu_{5,C} \cdot e_5 + \nu_{6,C} \cdot f_5 + \nu_{7,C} \cdot g_5) = 0$$

$$F_{22} \cdot x_{9,D} - F_{22} \cdot x_{4,D} + y3 \cdot (\nu_{1,D} \cdot a_5$$

$$+\nu_{2,D} \cdot b_5 + \nu_{3,D} \cdot c_5 + \nu_{4,D} \cdot d_5 + \nu_{5,D} \cdot e_5 + \nu_{6,D} \cdot f_5 + \nu_{7,D} \cdot g_5) = 0$$

$$F_{22} \cdot x_{9,E} - F_{22} \cdot x_{4,E} + y3 \cdot (\nu_{1,E} \cdot a_5$$

$$+\nu_{2,E} \cdot b_5 + \nu_{3,E} \cdot c_5 + \nu_{4,E} \cdot d_5 + \nu_{5,E} \cdot e_5 + \nu_{6,E} \cdot f_5 + \nu_{7,E} \cdot g_5) = 0$$

$$a_4 - k_1 \cdot x_{9,A} = 0$$

$$b_4 - k_2 \cdot x_{9,A} = 0$$

$$c_4 - k_3 \cdot x_{9,D} = 0$$

$$d_4 - k_4 \cdot x_{9,D}^2 = 0$$

$$e_4 - k_5 \cdot x_{9,A}^2 = 0$$

$$f_4 - k_6 \cdot x_{9,B} = 0$$

$$g_4 - k_7 \cdot x_{9,C} = 0$$

$$a_5 - k_1 \cdot x_{4,A} = 0$$

$$b_5 - k_2 \cdot x_{4,A} = 0$$

$$c_5 - k_3 \cdot x_{4,D} = 0$$

$$d_5 - k_4 \cdot x_{4,D}^2 = 0$$

$$e_5 - k_5 \cdot x_{4,A}^2 = 0$$

$$f_5 - k_6 \cdot x_{4,B} = 0$$

$$g_5 - k_7 \cdot x_{4,C} = 0$$

9.4.2 Data

- inlet compositions: l_m

	l_m
A	1.0
B	0
C	0
D	0
E	0

- stoichiometric coefficients: $\nu_{rp,m}$

	A	B	C	D	E
1	-1	0	0	0	1
2	-1	0	0	1	0
3	0	1	0	-1	0
4	0	0	1	-2	0
5	-2	0	1	0	0
6	0	-1	0	1	0
7	0	0	-1	2	0

- kinetic constants: k_i

	k_i
1	0.333840
2	0.266870
3	0.149400
4	0.189570
5	0.009598
6	0.294250
7	0.011932

- prespecified streams

	Value
F_1	100
F_{12}	100

- Upper and lower bounds

Variable		*Lower*	*Upper*
F_i	$i = 2, 3, 17, 18$	0.0	100
F_i	$i = 4, \ldots, 16, 19, \ldots, 22$	0.0	5,000
$x_{i,m}$	$m = A, B, C, D, E \quad i = 1, \ldots, 9$	0.0	3.0
$x_{5,D}$		0.01	3.0
$y1$		1	5,000
$y3$		0.001	10.0

9.4.3 Problem Statistics

Number of Variables	105
Number of Linear Constraints	38
Number of Nonlinear Constraints	60

9.4.4 Best Known Solution

- Objective value=1.402

- Flowrates

Var.	Level	Var.	Level	Var.	Level	Var.	Level
F_1	100	F_2	0.0	F_3	100.0	F_4	10.775
F_5	0.0	F_6	0.0	F_7	0.0	F_8	169.084
F_9	10.775	F_{10}	69.084	F_{11}	10.775	F_{12}	58.309
F_{13}	0.0	F_{14}	0.0	F_{15}	10.775	F_{16}	0.0
F_{17}	0.0	F_{18}	100.0	F_{19}	58.309	F_{20}	0.0
F_{21}	69.084	F_{22}	169.084	F_{23}	69.084		

- Concentrations

Var.	Level	Var.	Level	Var.	Level	Var.	Level
$x_{1,A}$	0.903	$x_{2,A}$	0.763	$x_{3,A}$	0.003	$x_{4,A}$	0.903
$x_{1,B}$	0.005	$x_{2,B}$	0.012	$x_{3,B}$	0.052	$x_{4,B}$	0.005
$x_{1,C}$	0.014	$x_{2,C}$	0.034	$x_{3,C}$	0.144	$x_{4,C}$	0.014
$x_{1,D}$	0.010	$x_{2,D}$	0.024	$x_{3,D}$	0.103	$x_{4,D}$	0.010
$x_{1,E}$	0.054	$x_{2,E}$	0.132	$x_{3,E}$	0.554	$x_{4,E}$	0.054
$x_{5,A}$	0.903	$x_{6,A}$	0.763	$x_{7,A}$	0.903	$x_{8,A}$	0.903
$x_{5,B}$	0.005	$x_{6,B}$	0.012	$x_{7,B}$	0.005	$x_{8,B}$	0.005
$x_{5,C}$	0.014	$x_{6,C}$	0.034	$x_{7,C}$	0.014	$x_{8,C}$	0.014
$x_{5,D}$	0.010	$x_{6,D}$	0.024	$x_{7,D}$	0.010	$x_{8,D}$	0.010
$x_{5,E}$	0.054	$x_{6,E}$	0.132	$x_{7,E}$	0.054	$x_{8,E}$	0.054
$x_{9,A}$	0.903	$x_{9,B}$	0.005	$x_{9,C}$	0.014	$x_{9,D}$	0.010
$x_{9,E}$	0.054						

- Reaction rates and volumes

Var.	Level	Var.	Level	Var.	Level	Var.	Level
a_1	0.001	b_1	8.61 E-4	c_1	0.015	d_1	0.002
a_2	0.255	b_2	0.204	c_2	0.004	d_2	1.3 E-4
a_3	0.255	b_3	0.204	c_3	0.004	d_3	1.13 E-4
a_4	0.301	b_4	0.241	c_4	0.001	d_4	1.89 E-5
a_5	0.301	b_5	0.241	c_5	0.001	d_5	1.89 E-5
e_1	0.000	f_1	0.015	g_1	0.002	$y1$	5,000
e_2	0.006	f_2	0.004	g_2	4.09 E-4	$y2$	0.004
e_3	0.006	f_3	0.004	g_3	4.09 E-4	$y3$	0.001
e_4	0.008	f_4	0.001	g_4	1.67 E-4		
e_5	0.008	f_5	0.001	g_5	1.67 E-4		

.5 Test Problem 4 : Denbigh reaction

he Denbigh reaction is considered in this example (see [125]) :

- *reaction mechanism:*

$$A \xrightarrow{k_1} B \quad (second\ order\ reaction\ ) \qquad (9.13)$$

$$B \xrightarrow{k_2} C \quad (first\ order\ reaction\ ) \qquad (9.14)$$

$$A \xrightarrow{k_3} D \quad (first\ order\ reaction\ ) \qquad (9.15)$$

$$B \xrightarrow{k_4} E \quad (second\ order\ reaction\ ) \qquad (9.16)$$

- *objective function:* Selectivity of B over D.

9.5.1 Problem Formulation

$$\text{MAX} \quad \frac{x_{5,B}}{x_{5,C}}$$

$$F_1 - F_2 - F_3 = 0$$
$$F_4 + F_7 - F_9 = 0$$
$$F_5 + F_6 + F_8 - F_{10} = 0$$

$$F_7 + F_8 - F_{23} - F_{12} = 0$$

$$F_7 \cdot x_{3,A} + F_8 \cdot x_{4,A} - F_{11} \cdot x_{5,A} - F_{12} \cdot x_{5,A} = 0$$

$$F_7 \cdot x_{3,B} + F_8 \cdot x_{4,B} - F_{11} \cdot x_{5,B} - F_{12} \cdot x_{5,B} = 0$$

$$F_7 \cdot x_{3,C} + F_8 \cdot x_{4,C} - F_{11} \cdot x_{5,C} - F_{12} \cdot x_{5,C} = 0$$

$$F_7 \cdot x_{3,D} + F_8 \cdot x_{4,D} - F_{11} \cdot x_{5,D} - F_{12} \cdot x_{5,D} = 0$$

$$F_7 \cdot x_{3,E} + F_8 \cdot x_{4,E} - F_{11} \cdot x_{5,E} - F_{12} \cdot x_{5,E} = 0$$

$$F_{23} - F_{11} - F_{12} = 0$$

$$F_5 + F_{11} + F_2 - F_9 = 0$$

$$F_5 \cdot x_{4,A} + F_{11} \cdot x_{5,A} + F_2 \cdot l_A - F_9 \cdot x_{1,A} = 0$$

$$F_5 \cdot x_{4,B} + F_{11} \cdot x_{5,B} + F_2 \cdot l_B - F_9 \cdot x_{1,B} = 0$$

$$F_5 \cdot x_{4,C} + F_{11} \cdot x_{5,C} + F_2 \cdot l_C - F_9 \cdot x_{1,C} = 0$$

$$F_5 \cdot x_{4,D} + F_{11} \cdot x_{5,D} + F_2 \cdot l_D - F_9 \cdot x_{1,D} = 0$$

$$F_5 \cdot x_{4,E} + F_{11} \cdot x_{5,E} + F_2 \cdot l_E - F_9 \cdot x_{1,E} = 0$$

$$F_9 \cdot x_{1,A} - F_9 \cdot x_{3,A} + y1 \cdot (\nu_{1,A} \cdot a_1 + \nu_{2,A} \cdot b_1 + \nu_{3,A} \cdot c_1 + \nu_{4,A} \cdot d_1) = 0$$

$$F_9 \cdot x_{1,B} - F_9 \cdot x_{3,B} + y1 \cdot (\nu_{1,B} \cdot a_1 + \nu_{2,B} \cdot b_1 + \nu_{3,B} \cdot c_1 + \nu_{4,B} \cdot d_1) = 0$$

$$F_9 \cdot x_{1,C} - F_9 \cdot x_{3,C} + y1 \cdot (\nu_{1,C} \cdot a_1 + \nu_{2,C} \cdot b_1 + \nu_{3,C} \cdot c_1 + \nu_{4,C} \cdot d_1) = 0$$

$$F_9 \cdot x_{1,D} - F_9 \cdot x_{3,D} + y1 \cdot (\nu_{1,D} \cdot a_1 + \nu_{2,D} \cdot b_1 + \nu_{3,D} \cdot c_1 + \nu_{4,D} \cdot d_1) = 0$$

$$F_9 \cdot x_{1,E} - F_9 \cdot x_{3,E} + y1 \cdot (\nu_{1,E} \cdot a_1 + \nu_{2,E} \cdot b_1 + \nu_{3,E} \cdot c_1 + \nu_{4,E} \cdot d_1) = 0$$

$$a_1 - k_1 \cdot x_{3,A}^2 = 0$$

$$b_1 - k_2 \cdot x_{3,A} = 0$$

$$c_1 - k_3 \cdot x_{3,B}^2 = 0$$

$$d_1 - k_4 \cdot x_{3,B} = 0$$

$$F_{13} - F_{21} + F_{16} + F_{18} + F_{20} - F_{14} - F_{22} = 0$$

$$F_{14} + F_{22} - F_{10} = 0$$

$$F_{13} \cdot x_{2,A} + F_{21} \cdot x_{7,A} + F_{16} \cdot x_{3,A} + F_{18} \cdot l_A$$
$$+ F_{20} \cdot x_{5,A} - F_{14} \cdot x_{8,A} - F_{22} \cdot x_{8,A} = 0$$

$$F_{13} \cdot x_{2,B} + F_{21} \cdot x_{7,B} + F_{16} \cdot x_{3,B} + F_{18} \cdot l_B$$

$$+F_{20} \cdot x_{5,B} - F_{14} \cdot x_{8,B} - F_{22} \cdot x_{8,B} = 0$$

$$F_{13} \cdot x_{2,C} + F_{21} \cdot x_{7,C} + F_{16} \cdot x_{3,C} + F_{18} \cdot l_C$$
$$+F_{20} \cdot x_{5,C} - F_{14} \cdot x_{8,C} - F_{22} \cdot x_{8,C} = 0$$

$$F_{13} \cdot x_{2,D} + F_{21} \cdot x_{7,D} + F_{16} \cdot x_{3,D} + F_{18} \cdot l_D$$
$$+F_{20} \cdot x_{5,D} - F_{14} \cdot x_{8,D} - F_{22} \cdot x_{8,D} = 0$$

$$F_{13} \cdot x_{2,E} + F_{21} \cdot x_{7,E} + F_{16} \cdot x_{3,E} + F_{18} \cdot l_E$$
$$+F_{20} \cdot x_{5,E} - F_{14} \cdot x_{8,E} - F_{22} \cdot x_{8,E} = 0$$

$$F_{14} \cdot x_{8,A} + F_{22} \cdot x_{4,A} - F_{10} \cdot x_{4,A} = 0$$
$$F_{14} \cdot x_{8,B} + F_{22} \cdot x_{4,B} - F_{10} \cdot x_{4,B} = 0$$
$$F_{14} \cdot x_{8,C} + F_{22} \cdot x_{4,C} - F_{10} \cdot x_{4,C} = 0$$
$$F_{14} \cdot x_{8,D} + F_{22} \cdot x_{4,D} - F_{10} \cdot x_{4,D} = 0$$
$$F_{14} \cdot x_{8,E} + F_{22} \cdot x_{4,E} - F_{10} \cdot x_{4,E} = 0$$

$$F_{15} + F_{17} + F_{19} + F_6 - F_{13} - F_{21} = 0$$

$$F_{15} \cdot x_{3,A} + F_{17} \cdot l_A + F_6 \cdot x_{4,A} + F_{19} \cdot x_{5,A} - F_{13} \cdot x_{2,A} - F_{21} \cdot x_{2,A} = 0$$
$$F_{15} \cdot x_{3,B} + F_{17} \cdot l_B + F_6 \cdot x_{4,B} + F_{19} \cdot x_{5,B} - F_{13} \cdot x_{2,B} - F_{21} \cdot x_{2,B} = 0$$
$$F_{15} \cdot x_{3,C} + F_{17} \cdot l_C + F_6 \cdot x_{4,C} + F_{19} \cdot x_{5,C} - F_{13} \cdot x_{2,C} - F_{21} \cdot x_{2,C} = 0$$
$$F_{15} \cdot x_{3,D} + F_{17} \cdot l_D + F_6 \cdot x_{4,D} + F_{19} \cdot x_{5,D} - F_{13} \cdot x_{2,D} - F_{21} \cdot x_{2,D} = 0$$
$$F_{15} \cdot x_{3,E} + F_{17} \cdot l_E + F_6 \cdot x_{4,E} + F_{19} \cdot x_{5,E} - F_{13} \cdot x_{2,E} - F_{21} \cdot x_{2,E} = 0$$

$$F_3 - F_{17} - F_{18} = 0$$
$$F_4 - F_{15} - F_{16} = 0$$
$$F_{12} - F_{19} - F_{20} = 0$$
$$y2 - 4 \cdot y3 = 0$$

$$F_{21} \cdot x_{2,A} - F_{21} \cdot x_{6,A} + y3 \cdot (\nu_{1,A} \cdot a_2 + \nu_{2,A} \cdot b_2 + \nu_{3,A} \cdot c_2 + \nu_{4,A} \cdot d_2) = 0$$
$$F_{21} \cdot x_{2,B} - F_{21} \cdot x_{6,B} + y3 \cdot (\nu_{1,B} \cdot a_2 + \nu_{2,B} \cdot b_2 + \nu_{3,B} \cdot c_2 + \nu_{4,B} \cdot d_2) = 0$$
$$F_{21} \cdot x_{2,C} - F_{21} \cdot x_{6,C} + y3 \cdot (\nu_{1,C} \cdot a_2 + \nu_{2,C} \cdot b_2 + \nu_{3,C} \cdot c_2 + \nu_{4,C} \cdot d_2) = 0$$
$$F_{21} \cdot x_{2,D} - F_{21} \cdot x_{6,D} + y3 \cdot (\nu_{1,D} \cdot a_2 + \nu_{2,D} \cdot b_2 + \nu_{3,D} \cdot c_2 + \nu_{4,D} \cdot d_2) = 0$$
$$F_{21} \cdot x_{2,E} - F_{21} \cdot x_{6,E} + y3 \cdot (\nu_{1,E} \cdot a_2 + \nu_{2,E} \cdot b_2 + \nu_{3,E} \cdot c_2 + \nu_{4,E} \cdot d_2) = 0$$

$$F_{21} \cdot x_{6,A} - F_{21} \cdot x_{7,A} + y3 \cdot (\nu_{1,A} \cdot a_3 + \nu_{2,A} \cdot b_3 + \nu_{3,A} \cdot c_3 + \nu_{4,A} \cdot d_3) = 0$$

$$F_{21} \cdot x_{6,B} - F_{21} \cdot x_{7,B} + y3 \cdot (\nu_{1,B} \cdot a_3 + \nu_{2,B} \cdot b_3 + \nu_{3,B} \cdot c_3 + \nu_{4,B} \cdot d_3) = 0$$

$$F_{21} \cdot x_{6,C} - F_{21} \cdot x_{7,C} + y3 \cdot (\nu_{1,C} \cdot a_3 + \nu_{2,C} \cdot b_3 + \nu_{3,C} \cdot c_3 + \nu_{4,C} \cdot d_3) = 0$$

$$F_{21} \cdot x_{6,D} - F_{21} \cdot x_{7,D} + y3 \cdot (\nu_{1,D} \cdot a_3 + \nu_{2,D} \cdot b_3 + \nu_{3,D} \cdot c_3 + \nu_{4,D} \cdot d_3) = 0$$

$$F_{21} \cdot x_{6,E} - F_{21} \cdot x_{7,E} + y3 \cdot (\nu_{1,E} \cdot a_3 + \nu_{2,E} \cdot b_3 + \nu_{3,E} \cdot c_3 + \nu_{4,E} \cdot d_3) = 0$$

$$a_2 - k_1 \cdot x_{6,A}^2 = 0$$

$$b_2 - k_2 \cdot x_{6,A} = 0$$

$$c_2 - k_3 \cdot x_{6,B}^2 = 0$$

$$d_2 - k_4 \cdot x_{6,B} = 0$$

$$a_3 - k_1 \cdot x_{7,A}^2 = 0$$

$$b_3 - k_2 \cdot x_{7,A} = 0$$

$$c_3 - k_3 \cdot x_{7,B}^2 = 0$$

$$d_3 - k_4 \cdot x_{7,B} = 0$$

$$F_{22} \cdot x_{8,A} - F_{22} \cdot x_{9,A} + y3 \cdot (\nu_{1,A} \cdot a_4 + \nu_{2,A} \cdot b_4 + \nu_{3,A} \cdot c_4 + \nu_{4,A} \cdot d_4) = 0$$

$$F_{22} \cdot x_{8,B} - F_{22} \cdot x_{9,B} + y3 \cdot (\nu_{1,B} \cdot a_4 + \nu_{2,B} \cdot b_4 + \nu_{3,B} \cdot c_4 + \nu_{4,B} \cdot d_4) = 0$$

$$F_{22} \cdot x_{8,C} - F_{22} \cdot x_{9,C} + y3 \cdot (\nu_{1,C} \cdot a_4 + \nu_{2,C} \cdot b_4 + \nu_{3,C} \cdot c_4 + \nu_{4,C} \cdot d_4) = 0$$

$$F_{22} \cdot x_{8,D} - F_{22} \cdot x_{9,D} + y3 \cdot (\nu_{1,D} \cdot a_4 + \nu_{2,D} \cdot b_4 + \nu_{3,D} \cdot c_4 + \nu_{4,D} \cdot d_4) = 0$$

$$F_{22} \cdot x_{8,E} - F_{22} \cdot x_{9,E} + y3 \cdot (\nu_{1,E} \cdot a_4 + \nu_{2,E} \cdot b_4 + \nu_{3,E} \cdot c_4 + \nu_{4,E} \cdot d_4) = 0$$

$$F_{22} \cdot x_{9,A} - F_{22} \cdot x_{4,A} + y3 \cdot (\nu_{1,A} \cdot a_5 + \nu_{2,A} \cdot b_5 + \nu_{3,A} \cdot c_5 + \nu_{4,A} \cdot d_5) = 0$$

$$F_{22} \cdot x_{9,B} - F_{22} \cdot x_{4,B} + y3 \cdot (\nu_{1,B} \cdot a_5 + \nu_{2,B} \cdot b_5 + \nu_{3,B} \cdot c_5 + \nu_{4,B} \cdot d_5) = 0$$

$$F_{22} \cdot x_{9,C} - F_{22} \cdot x_{4,C} + y3 \cdot (\nu_{1,C} \cdot a_5 + \nu_{2,C} \cdot b_5 + \nu_{3,C} \cdot c_5 + \nu_{4,C} \cdot d_5) = 0$$

$$F_{22} \cdot x_{9,D} - F_{22} \cdot x_{4,D} + y3 \cdot (\nu_{1,D} \cdot a_5 + \nu_{2,D} \cdot b_5 + \nu_{3,D} \cdot c_5 + \nu_{4,D} \cdot d_5) = 0$$

$$F_{22} \cdot x_{9,E} - F_{22} \cdot x_{4,E} + y3 \cdot (\nu_{1,E} \cdot a_5 + \nu_{2,E} \cdot b_5 + \nu_{3,E} \cdot c_5 + \nu_{4,E} \cdot d_5) = 0$$

$$a_4 - k_1 \cdot x_{9,A}^2 = 0$$

$$b_4 - k_2 \cdot x_{9,A} = 0$$

$$c_4 - k_3 \cdot x_{9,B}^2 = 0$$

$$d_4 - k_4 \cdot x_{9,B} = 0$$

$$a_5 - k_1 \cdot x_{4,A}^2 = 0$$

$$b_5 - k_2 \cdot x_{4,A} = 0$$

$$c_5 - k_3 \cdot x_{4,B}^2 = 0$$

$$d_5 - k_4 \cdot x_{4,B} = 0$$

9.5.2 Data

- inlet compositions: l_m

	l_m
A	5.8
B	6.0
C	0
D	0.6
E	0

- stoichiometric coefficients: $\nu_{rp,m}$

	A	B	C	D	E
1	-1	0.5	0	0	0
2	-1	0	1	0	0
3	0	-1	0	0	1
4	0	-1	0	1	0

- kinetic constants: k_i

	k_i
1	1.0
2	0.6
3	0.1
4	0.6

- prespecified streams

	Value
F_1	100
F_{12}	100

– Upper and lower bounds

Variable		Lower	Upper
F_i	$i = 2, 3, 17, 18$	0.0	100
F_i	$i = 4, \ldots, 16, 19, \ldots, 22$	0.0	5,000
$x_{i,m}$	$m = A, B, C, D, E \quad i = 1, \ldots, 9$	0.0	6.0
$y1$		10.0	500.00
$y3$		0.1	70.00

9.5.3 Problem Statistics

Number of Variables	90
Number of Linear Constraints	23
Number of Nonlinear Constraints	60

9.5.4 Best Known Solution

– Objective value=1.195

– Flowrates

$$F_i = 100 \qquad i = 1, 2, 4, 8, 9, 10, 15, 21, 22$$

– Compositions

Var.	Level	Var.	Level	Var.	Level	Var.	Level
$x_{1,A}$	6.0	$x_{2,A}$	4.086	$x_{3,A}$	4.086	$x_{4,A}$	2.579
$x_{1,B}$	0.0	$x_{2,B}$	0.782	$x_{3,B}$	0.782	$x_{4,B}$	1.302
$x_{1,C}$	0.6	$x_{2,C}$	0.845	$x_{3,C}$	0.845	$x_{4,C}$	1.090
$x_{1,D}$	0.0	$x_{2,D}$	0.047	$x_{3,D}$	0.047	$x_{4,D}$	0.139
$x_{1,E}$	0.0	$x_{2,E}$	0.006	$x_{3,E}$	0.006	$x_{4,E}$	0.024
$x_{5,A}$	2.579	$x_{6,A}$	3.584	$x_{7,A}$	3.182	$x_{8,A}$	3.182
$x_{5,B}$	1.302	$x_{6,B}$	0.974	$x_{7,B}$	1.116	$x_{8,B}$	1.116
$x_{5,C}$	1.090	$x_{6,C}$	0.957	$x_{7,C}$	0.981	$x_{8,C}$	0.981
$x_{5,D}$	0.139	$x_{6,D}$	0.066	$x_{7,D}$	0.089	$x_{8,D}$	0.089
$x_{5,E}$	0.024	$x_{6,E}$	0.009	$x_{7,E}$	0.013	$x_{8,E}$	0.013
$x_{9,A}$	2.853	$x_{9,B}$	1.223	$x_{9,C}$	1.038	$x_{9,D}$	0.013
$x_{9,E}$	0.018						

- Reaction rates and volumes

Var.	Level	Var.	Level	Var.	Level	Var.	Level
a_1	16.692	b_1	2.451	c_1	0.061	d_1	0.469
a_2	12.848	b_2	2.151	c_2	0.095	d_2	0.584
a_3	10.127	b_3	1.909	c_3	0.125	d_3	0.670
a_4	8.140	b_4	1.712	c_4	0.150	d_4	0.734
a_5	6.651	b_5	1.547	c_5	0.170	d_5	0.781
$y1$	10.0	$y2$	13.367	$y3$	3.342		

9.6 Test Problem 5 : Levenspiel reaction

In this example, the interesting case of an autocatalytic reaction was considered, in which the reaction rate increases with increasing conversion up to a point. This test problem is taken from [125].

- *reaction mechanism:*

- $A + B \xrightarrow{k} 2\,B$ (*second order reaction*)

- *objective function*: Maximization of yield of B

9.6.1 Problem Formulation

$$\text{MAX} \quad x_{5,B}$$

$$F_1 - F_2 - F_3 = 0$$

$$F_4 + F_7 - F_9 = 0$$

$$F_5 + F_6 + F_8 - F_{10} = 0$$

$$F_7 + F_8 - F_{23} - F_{12} = 0$$

$$F_7 \cdot x_{3,A} + F_8 \cdot x_{4,A} - F_{11} \cdot x_{5,A} - F_{12} \cdot x_{5,A} = 0$$

$$F_7 \cdot x_{3,B} + F_8 \cdot x_{4,B} - F_{11} \cdot x_{5,B} - F_{12} \cdot x_{5,B} = 0$$

$$F_{23} - F_{11} - F_{12} = 0$$

$$F_5 + F_{11} + F_2 - F_9 = 0$$

$$F_5 \cdot x_{4,A} + F_{11} \cdot x_{5,A} + F_2 \cdot l_A - F_9 \cdot x_{1,A} = 0$$

$$F_5 \cdot x_{4,B} + F_{11} \cdot x_{5,B} + F_2 \cdot l_B - F_9 \cdot x_{1,B} = 0$$

$$F_9 \cdot x_{1,A} - F_9 \cdot x_{3,A} + \nu_{1,A} \cdot y1 \cdot a1 = 0$$

$$F_9 \cdot x_{1,B} - F_9 \cdot x_{3,B} + \nu_{1,B} \cdot y1 \cdot a1 = 0$$

$$a1 - k1 \cdot x_{3,A} \cdot x_{3,B} = 0$$

$$F_{13} - F_{21} + F_{16} + F_{18} + F_{20} - F_{14} - F_{22} = 0$$

$$F_{14} + F_{22} - F_{10} = 0$$

$$F_{13} \cdot x_{2,A} + F_{21} \cdot x_{7,A} + F_{16} \cdot x_{3,A} + F_{18} \cdot l_A$$
$$+ F_{20} \cdot x_{5,A} - F_{14} \cdot x_{8,A} - F_{22} \cdot x_{8,A} = 0$$

$$F_{13} \cdot x_{2,B} + F_{21} \cdot x_{7,B} + F_{16} \cdot x_{3,B} + F_{18} \cdot l_B$$
$$+ F_{20} \cdot x_{5,B} - F_{14} \cdot x_{8,B} - F_{22} \cdot x_{8,B} = 0$$

$$F_{14} \cdot x_{8,A} + F_{22} \cdot x_{4,A} - F_{10} \cdot x_{4,A} = 0$$

$$F_{14} \cdot x_{8,B} + F_{22} \cdot x_{4,B} - F_{10} \cdot x_{4,B} = 0$$

$$F_{15} + F_{17} + F_{19} + F_6 - F_{13} - F_{21} = 0$$

$$F_{15} \cdot x_{3,A} + F_{17} \cdot l_A + F_6 \cdot x_{4,A} + F_{19} \cdot x_{5,A} - F_{13} \cdot x_{2,A} - F_{21} \cdot x_{2,A} = 0$$

$$F_{15} \cdot x_{3,B} + F_{17} \cdot l_B + F_6 \cdot x_{4,B} + F_{19} \cdot x_{5,B} - F_{13} \cdot x_{2,B} - F_{21} \cdot x_{2,B} = 0$$

$$F_3 - F_{17} - F_{18} = 0$$
$$F_4 - F_{15} - F_{16} = 0$$
$$F_{12} - F_{19} - F_{20} = 0$$
$$y2 - 4 \cdot y3 = 0$$
$$F_{21} \cdot x_{2,A} - F_{21} \cdot x_{6,A} + \nu_{1,A} \cdot y3 \cdot a2 = 0$$
$$F_{21} \cdot x_{2,B} - F_{21} \cdot x_{6,B} + \nu_{1,B} \cdot y3 \cdot a2 = 0$$
$$F_{21} \cdot x_{6,A} - F_{21} \cdot x_{7,A} + \nu_{1,A} \cdot y3 \cdot a3 = 0$$
$$F_{21} \cdot x_{6,B} - F_{21} \cdot x_{7,B} + \nu_{1,B} \cdot y3 \cdot a3 = 0$$
$$a2 - k1 \cdot x_{6,A} \cdot x_{6,B} = 0$$
$$a3 - k1 \cdot x_{7,A} \cdot x_{7,B} = 0$$
$$F_{22} \cdot x_{8,A} - F_{22} \cdot x_{9,A} + \nu_{1,A} \cdot y3 \cdot a4 = 0$$
$$F_{22} \cdot x_{8,B} - F_{22} \cdot x_{9,B} + \nu_{1,B} \cdot y3 \cdot a4 = 0$$
$$F_{22} \cdot x_{9,A} - F_{22} \cdot x_{4,A} + \nu_{1,A} \cdot y3 \cdot a5 = 0$$
$$F_{22} \cdot x_{9,B} - F_{22} \cdot x_{4,B} + \nu_{1,B} \cdot y3 \cdot a5 = 0$$
$$a4 - k1 \cdot x_{9,A} \cdot x_{9,B} = 0$$
$$a5 - k1 \cdot x_{4,A} \cdot x_{4,B} = 0$$

9.6.2 Data

- inlet compositions: l_m

	l_m
A	0.44
B	0.55

- stoichiometric coefficients: $\nu_{rp,m}$

	A	B
1	-1	1

- kinetic constants: k_i

	k_i
1	1.0

– prespecified streams

	Value
F_1	100
F_{12}	100

– Upper and lower bounds

Variable		Lower	Upper
F_i	$i = 2, 3, 17, 18$	0.0	100
F_i	$i = 4, \ldots, 16, 19, \ldots, 22$	0.0	5,000
$x_{i,A}$	$i = 1, \ldots, 9$	0.0	0.45
$x_{i,B}$	$i = 1, \ldots, 9$	0.55	1.00

9.6.3   Problem Statistics

Number of Variables	48
Number of Linear Constraints	13
Number of Nonlinear Constraints	25

9.6.4   Best Known Solution

– Objective value=0.761

– Flowrates

$$F_i = 100 \qquad i = 1, 2, 4, 8, 9, 10, 15, 21, 22$$

– Compositions

Var.	Level	Var.	Level	Var.	Level	Var.	Level
$x_{1,A}$	0.450	$x_{2,A}$	0.338	$x_{3,A}$	0.338	$x_{4,A}$	0.239
$x_{1,B}$	0.550	$x_{2,B}$	0.662	$x_{3,B}$	0.662	$x_{4,B}$	0.761
$x_{5,A}$	0.239	$x_{6,A}$	0.311	$x_{7,A}$	0.286	$x_{8,A}$	0.286
$x_{5,B}$	0.761	$x_{6,B}$	0.689	$x_{7,B}$	0.738	$x_{8,B}$	0.738
$x_{9,A}$	0.262	$x_{9,B}$	0.761				

– Reaction rates

Var.	Level	Var.	Level	Var.	Level
a_1	0.224	a_2	0.214	a_3	0.204
a_4	0.193	a_5	0.182		

Chapter 10

Reactor-Separator-Recycle System test problems

10.1 Problem Statement

In most chemical processes reactors are sequenced by systems that separate the desired products out of their outlet reactor streams and recycle the unconverted reactants back to the reactor system. The reactor-separator-recycle optimization problem can be stated as follows :

> *For a given chemical process in which a reaction system of known kinetics is followed by a sequence of separation tasks that are required to extract the desired products and recycle the unconverted reactants, the optimization problem consists of systematically determining the reactor/separator/recycle system that operates so that a given performance criterion (e.g. total cost or profit of the plant, yield or selectivity of desired products, conversion of reactants) is optimized. The solution of such a problem should provide information about:*
>
> **a.** *the reactor network (types and sizes of reactors, feeding strategy and interconnections among the reactors) ;*

---

b. *the separator network (appropriate separation sequence and sizes of separators) ;*

c. *the interconnection between the two networks via the allocations of the outlet streams from the reactors and the allocations of the recycles from the separators back to the reactors.*

The mathematical formulation of the problem for a fixed number of reactors and separators results in a nonconvex nonlinear programming NLP problem and is described in [124].

10.2 Test Problem 1 : Benzene Chlorination system

The design of a benzene chlorination process is considered as a first example which is taken from [124].

− *reaction process:*

$$C_6H_6 + Cl_2 \xrightarrow{k_1} C_6H_5Cl + HCl \qquad (10.1)$$
$$C_6H_5Cl + Cl_2 \xrightarrow{k_2} C_6H_4Cl_2 + HCl \qquad (10.2)$$

− *separation process:* the components to be separated consist of benzene (A), monochloro benzene (B) and dichlorobenzene (C);

− *recycled components:* Benzene (A)

Further chlorination reactions can also take place but since they involve insignificant amounts of reactants they have been considered to be negligible. The chlorination of benzene(A), monochlorobenzene(B) and dichlorobenzene(C) is in all cases first-order and irreversible.

10.2.1 Problem Formulation

$$\text{MIN} \quad s_1 + s_2 \cdot y1 + s_3 \cdot y2 + F_{14} \cdot (s_4 \cdot x_{11,A} + s_5 \cdot x_{11,B})$$
$$+ F_{15} \cdot (s_6 \cdot x_{12,B} + s_7 \cdot x_{12,B}^2) + s_8 \cdot (q_1 + q_2)$$

$$F_1 - F_2 - F_3 = 0$$

$$F_4 + F_7 - F_9 = 0$$

$$F_5 + F_6 + F_8 - F_{10} = 0$$

$$F_3 - F_{20} - F_{21} - F_{22} = 0$$

$$F_4 - F_{32} - F_{33} - F_{34} = 0$$

$$F_{12} - F_{26} - F_{27} - F_{28} = 0$$

$$F_5 + F_{11} + F_2 - F_9 = 0$$

$$F_5 \cdot x_{4,A} + F_2 + F_{11} \cdot x_{13,A} - F_9 \cdot x_{1,A} = 0$$

$$F_5 \cdot x_{4,B} + F_{11} \cdot x_{13,B} - F_9 \cdot x_{1,B} = 0$$

$$F_6 + F_{29} + F_{23} + F_{33} + F_{21} + F_{27} - F_{30} - F_{24} = 0$$

$$F_{30} + F_{24} + F_{34} + F_{22} + F_{28} - F_{31} - F_{25} = 0$$

$$F_{31} + F_{25} - F_{10} = 0$$

$$F_{29} \cdot x_{5,A} + F_{23} \cdot x_{6,A} + F_{33} \cdot x_{3,A} + F_{21} + F_{27} \cdot x_{13,A} + F_6 \cdot x_{4,A}$$
$$- F_{30} \cdot x_{7,A} - F_{24} \cdot x_{7,A} = 0$$

$$F_{29} \cdot x_{5,B} + F_{23} \cdot x_{6,B} + F_{33} \cdot x_{3,B} + F_{27} \cdot x_{13,B} + F_6 \cdot x_{4,B}$$
$$- F_{30} \cdot x_{7,B} - F_{24} \cdot x_{7,B} = 0$$

$$F_{30} \cdot x_{7,A} + F_{24} \cdot x_{8,A} + F_{34} \cdot x_{3,A} + F_{22} + F_{28} \cdot x_{13,A}$$
$$- F_{31} \cdot x_{9,A} - F_{25} \cdot x_{9,A} = 0$$

$$F_{30} \cdot x_{7,B} + F_{24} \cdot x_{8,B} + F_{34} \cdot x_{3,B} + F_{28} \cdot x_{13,B}$$
$$- F_{31} \cdot x_{9,B} - F_{25} \cdot x_{9,B} = 0$$

$$F_{31} \cdot x_{9,A} + F_{25} \cdot x_{10,A} - F_{10} \cdot x_{4,A} = 0$$

$$F_{31} \cdot x_{9,B} + F_{25} \cdot x_{10,B} - F_{10} \cdot x_{4,B} = 0$$

$$F_{32} + F_{20} + F_{26} - F_{29} - F_{23} = 0$$

$$F_{32} \cdot x_{3,A} + F_{20} + F_{26} \cdot x_{13,A} - F_{23} \cdot x_{5,A} = 0$$

$$F_{32} \cdot x_{3,B} + F_{26} \cdot x_{13,B} - F_{23} \cdot x_{5,B} = 0$$

$$y2 - 3 \cdot y3 = 0$$

$$F_9 \cdot x_{1,A} - F_9 \cdot x_{3,A} + y1 \cdot (\nu_{1,A} \cdot a_1 + \nu_{2,A} \cdot b_1) = 0$$

$$F_9 \cdot x_{1,B} - F_9 \cdot x_{3,B} + y1 \cdot (\nu_{1,B} \cdot a_1 + \nu_{2,B} \cdot b_1) = 0$$

$$F_{23} \cdot x_{5,A} - F_{23} \cdot x_{6,A} + y3 \cdot (\nu_{1,A} \cdot a_2 + \nu_{2,A} \cdot b_2) = 0$$

$$F_{23} \cdot x_{5,B} - F_{23} \cdot x_{6,B} + y3 \cdot (\nu_{1,B} \cdot a_2 + \nu_{2,B} \cdot b_2) = 0$$

$$F_{24} \cdot x_{7,A} - F_{24} \cdot x_{8,A} + y3 \cdot (\nu_{1,A} \cdot a_3 + \nu_{2,A} \cdot b_3) = 0$$

$$F_{24} \cdot x_{7,B} - F_{24} \cdot x_{8,B} + y3 \cdot (\nu_{1,B} \cdot a_3 + \nu_{2,B} \cdot b_3) = 0$$

$$F_{25} \cdot x_{9,A} - F_{25} \cdot x_{10,A} + y3 \cdot (\nu_{1,A} \cdot a_4 + \nu_{2,A} \cdot b_4) = 0$$

$$F_{25} \cdot x_{9,B} - F_{25} \cdot x_{10,B} + y3 \cdot (\nu_{1,B} \cdot a_4 + \nu_{2,B} \cdot b_4) = 0$$

$$a_1 - k_1 \cdot z_{1,A} = 0$$

$$a_2 - k_1 \cdot z_{2,A} = 0$$

$$a_3 - k_1 \cdot z_{3,A} = 0$$

$$a_4 - k_1 \cdot z_{4,A} = 0$$

$$b_1 - k_2 \cdot z_{1,B} = 0$$

$$b_2 - k_2 \cdot z_{2,B} = 0$$

$$b_3 - k_2 \cdot z_{3,B} = 0$$

$$b_4 - k_2 \cdot z_{4,B} = 0$$

$$F_7 + F_8 - F_{14} - F_{19} = 0$$

$$F_7 \cdot x_{3,A} + F_8 \cdot x_{4,A} - F_{14} \cdot x_{11,A} - F_{19} \cdot x_{11,A} = 0$$

$$F_7 \cdot x_{3,B} + F_8 \cdot x_{4,B} - F_{14} \cdot x_{11,B} - F_{19} \cdot x_{11,B} = 0$$

$$F_{14} - F_{15} - F_{17} = 0$$

$$x_{12,B} \cdot (x_{11,B} + x_{11,C}) - x_{11,B} = 0$$

$$F_{17} - F_{14} \cdot x_{11,A} = 0$$

$$F_{15} - F_{18} - F_{16} = 0$$

$$F_{18} - x_{12,B} \cdot F_{15} = 0$$

$$F_{11} + F_{12} - F_{19} - F_{17} + F_{13} = 0$$

$$F_{11} \cdot x_{13,A} + F_{12} \cdot x_{13,A} - F_{19} \cdot x_{11,A} - F_{17} \cdot x_{11,A} + F_{13} \cdot x_{13,A} = 0$$

$$F_{11} \cdot x_{13,B} + F_{12} \cdot x_{13,B} - F_{19} \cdot x_{11,B} + F_{13} \cdot x_{13,B} = 0$$

$$x_{1,A} + x_{1,B} + x_{1,C} - 1 = 0$$

$$x_{2,A} + x_{2,B} + x_{2,C} - 1 = 0$$

$$x_{3,A} + x_{3,B} + x_{3,C} - 1 = 0$$

$$x_{4,A} + x_{4,B} + x_{4,C} - 1 = 0$$

$$x_{5,A} + x_{5,B} + x_{5,C} - 1 = 0$$

$$x_{6,A} + x_{6,B} + x_{6,C} - 1 = 0$$

$$x_{7,A} + x_{7,B} + x_{7,C} - 1 = 0$$

$$x_{8,A} + x_{8,B} + x_{8,C} - 1 = 0$$

$$x_{9,A} + x_{9,B} + x_{9,C} - 1 = 0$$

$$x_{10,A} + x_{10,B} + x_{10,C} - 1 = 0$$

$$x_{11,A} + x_{11,B} + x_{11,C} - 1 = 0$$

$$x_{12,B} + x_{12,C} - 1 = 0$$

$$x_{13,A} + x_{13,B} + x_{13,C} - 1 = 0$$

$$v_1 - p_1 \cdot x_{3,A} + p_2 \cdot x_{3,B} + p_3 \cdot x_{3,C} = 0$$

$$v_2 - p_1 \cdot x_{6,A} + p_2 \cdot x_{6,B} + p_3 \cdot x_{6,C} = 0$$

$$v_3 - p_1 \cdot x_{8,A} + p_2 \cdot x_{8,B} + p_3 \cdot x_{8,C} = 0$$

$$v_4 - p_1 \cdot x_{10,A} + p_2 \cdot x_{10,B} + p_3 \cdot x_{10,C} = 0$$

$$z_{1,A} \cdot v_1 - x_{3,A} = 0$$

$$z_{2,A} \cdot v_2 - x_{6,A} = 0$$

$$z_{3,A} \cdot v_3 - x_{8,A} = 0$$

$$z_{4,A} \cdot v_4 - x_{10,A} = 0$$

$$z_{1,B} \cdot v_1 - x_{3,B} = 0$$

$$z_{2,B} \cdot v_2 - x_{6,B} = 0$$

$$z_{3,B} \cdot v_3 - x_{8,B} = 0$$

$$z_{4,B} \cdot v_4 - x_{10,B} = 0$$

$$z_{1,C} \cdot v_1 - x_{3,C} = 0$$

$$z_{2,C} \cdot v_2 - x_{6,C} = 0$$

$$z_{3,C} \cdot v_3 - x_{8,C} = 0$$
$$z_{4,C} \cdot v_4 - x_{10,C} = 0$$
$$F_{18} - 50 \geq 0$$
$$q_1 - F_{14} \cdot (s_9 + s_{10} \cdot x_{11,A} + s_{11} \cdot x_{11,B}) = 0$$
$$q_2 - F_{15} \cdot (s_{12} + s_{13} \cdot x_{12,B}) = 0$$

10.2.2 Data

− stoichiometric coefficients: $\nu_{rp,m}$

	A	B	C
1	-1	1	0
2	0	-1	1

− kinetic constants: k_i

	k_i
1	0.412
2	0.055

− cost parameters

Parameter	Value	Parameter	Value
s_1	10,317.702	s_8	19,0114
s_2	3,271.2252	s_9	3.003
s_3	19,729.086	s_{10}	36.106
s_4	147.62	s_{11}	7.706
s_5	-445.544	s_{12}	26.212
s_6	2,793.792	s_{13}	29.447
s_7	-1,547.812		

− molecular volumes p_i

	p_i
1	0.088
2	0.102
3	0.114

– Upper and lower bounds

	Variable		Lower	Upper
F_i	$i = 2, 3, 20, 21, 22$		0.0	100
F_i	$i = 4, \ldots, 19, 20, \ldots, 34$		0.0	500
$x_{i,m}$	$m = A, B, C$	$i = 1, \ldots, 13$	0.0	1.0
$z_{i,A}$	$i = 1, \ldots, 4$		0.0	11.031
$z_{i,B}$	$i = 1, \ldots, 4$		0.0	6.443
$z_{i,C}$	$i = 1, \ldots, 4$		0.0	0.531
$y1$			0.1	50
$y3$			0.1	4.0

10.2.3 Problem Statistics

Number of Variables	102
Number of Linear Constraints	38
Number of Nonlinear Constraints	43

10.2.4 Best Known Solution

– Objective value=339,833.412

– Flowrates

Var.	Level	Var.	Level	Var.	Level	Var.	Level
F_1	52.707	F_2	0.0	F_3	52.707	F_4	0.0
F_5	166.521	F_6	73.591	F_7	166.521	F_8	0.0
F_9	166.521	F_{10}	218.284	F_{11}	0.0	F_{12}	113.814
F_{13}	0.0	F_{14}	166.521	F_{15}	0.0	F_{16}	2.707
F_{17}	113.814	F_{18}	50.0	F_{19}	218.284	F_{20}	52.707
F_{21}	0.0	F_{22}	91.986	F_{23}	21.828	F_{24}	240.113
F_{25}	240.113	F_{26}	0.0	F_{27}	0.0	F_{28}	0.0
F_{29}	0.0	F_{30}	0.0	F_{31}	0.0	F_{32}	73.591
F_{33}	0.0	F_{34}	0.0				

- Concentrations and molar fractions

Var.	Level	Var.	Level	Var.	Level	Var.	Level
$z_{1,A}$	7.379	$z_{2,A}$	11.031	$z_{3,A}$	10.392	$z_{4,A}$	10.805
$z_{1,B}$	3.242	$z_{2,B}$	0.286	$z_{3,B}$	0.370	$z_{4,B}$	0.479
$z_{1,C}$	0.176	$z_{2,C}$	0.001	$z_{3,C}$	0.002	$z_{4,C}$	0.002
$x_{1,A}$	0.957	$x_{2,A}$	0.986	$x_{3,A}$	0.683	$x_{4,A}$	0.957
$x_{1,B}$	0.042	$x_{2,B}$	0.014	$x_{3,B}$	0.300	$x_{4,B}$	0.042
$x_{1,C}$	1.97 E-4	$x_{2,C}$	6.65 E-4	$x_{3,C}$	0.016	$x_{4,C}$	1.97 E-4
$x_{5,A}$	0.986	$x_{6,A}$	0.957	$x_{7,A}$	0.977	$x_{8,A}$	0.967
$x_{5,B}$	0.014	$x_{6,B}$	0.025	$x_{7,B}$	0.023	$x_{8,B}$	0.033
$x_{5,C}$	6.65 E-4	$x_{6,C}$	1.04 E-4	$x_{7,C}$	9.49 E-5	$x_{8,C}$	1.39 E-4
$x_{9,A}$	0.967	$x_{10,A}$	0.957	$x_{11,A}$	0.683	$x_{12,A}$	0.00
$x_{9,B}$	0.033	$x_{10,B}$	0.042	$x_{11,B}$	0.300	$x_{12,B}$	0.949
$x_{9,C}$	1.39 E-4	$x_{10,C}$	1.97 E-4	$x_{11,C}$	0.016	$x_{12,C}$	0.051
$x_{13,A}$	1.000	$x_{13,B}$	0.00	$x_{13,C}$	0.00		

- Reaction rates, volumes and molecular volumes

Var.	Level	Var.	Level	Var.	Level	Var.	Level
a_1	3.040	a_2	4.545	a_3	4.504	a_4	4.452
b_1	0.178	b_2	0.016	b_3	0.020	b_4	0.026
v_1	0.093	v_2	0.088	v_3	0.088	v_4	0.089
q_1	4994.735	q_2	2853.912	y1	15.0	y2	1.579
y3	0.526						

10.3 Test Problem 2 : Production of Ethylbenzene system

In this example, which is taken from [124], the alkylation of benzene with ethylene for the production of ethylbenzene was studied. The process is an intermediate stage of the production of styrene using the *direct hydrogenation method* and is carried out in liquid phase. The following first order reversible reactions were found to describe this process:

- *reaction process*:

$$C_6H_6 + CH_2 = CH_2 \overset{k_1}{\rightleftharpoons} C_6H_5CH_2CH_3 \qquad (10.3)$$

$$C_6H_5CH_2CH_3 + CH_2 = CH_2 \overset{k_2}{\rightleftharpoons} C_6H_4(CH_2CH_3)_2 \qquad (10.4)$$

In the alkylators liquid benzene (A) reacts with a gaseous stream of pure ethylene to produce the desired ethylbenzene (B) and the co-product diethylbenzene (C). In the separation level, B is obtained at a minimum rate of 10 $kmol/hr$ while both A and C have potential recycles to the reactor network.

- *separation process*: the components to be separated consist of benzene (A), ethylbenzene (B) and diethylbenzene (C);

- *recycled components*: benzene (A), diethylbenzene (C)

10.3.1 Problem Formulation

$$\text{MIN} \quad s_1 + s_2 \cdot (y1 + y2) + s_3 \cdot (r_1 + r_2)$$

$$F_1 - F_2 - F_3 = 0$$
$$F_4 + F_7 - F_9 = 0$$
$$F_5 + F_6 + F_8 - F_{10} = 0$$
$$F_3 - F_{20} - F_{21} - F_{22} = 0$$
$$F_4 - F_{32} - F_{33} - F_{34} = 0$$
$$F_{12} - F_{26} - F_{27} - F_{28} = 0$$
$$F_{40} - F_{36} - F_{37} - F_{38} = 0$$
$$F_5 + F_{11} + F_{39} + F_2 - F_9 = 0$$
$$F_5 \cdot x_{4,A} + F_2 + F_{11} \cdot x_{13,A} + F_{39} \cdot x_{14,A} - F_9 \cdot x_{1,A} = 0$$
$$F_5 \cdot x_{4,B} + F_{11} \cdot x_{13,B} + F_{39} \cdot x_{14,B} - F_9 \cdot x_{1,B} = 0$$

$$F_6 + F_{29} + F_{23} + F_{33} + F_{21} + F_{27} + F_{37} - F_{30} - F_{24} = 0$$

$$F_{30} + F_{24} + F_{34} + F_{22} + F_{28} + F_{38} - F_{31} - F_{25} = 0$$

$$F_{31} + F_{25} - F_{10} = 0$$

$$F_{29} \cdot x_{5,A} + F_{23} \cdot x_{6,A} + F_{33} \cdot x_{3,A} + F_{21} + F_{27} \cdot x_{13,A}$$
$$+ F_{37} \cdot x_{14,A} + F_6 \cdot x_{4,A} - F_{30} \cdot x_{7,A} - F_{24} \cdot x_{7,A} = 0$$

$$F_{29} \cdot x_{5,B} + F_{23} \cdot x_{6,B} + F_{33} \cdot x_{3,B} + F_{27} \cdot x_{13,B}$$
$$+ F_{37} \cdot x_{14,B} + F_6 \cdot x_{4,B} - F_{30} \cdot x_{7,B} - F_{24} \cdot x_{7,B} = 0$$

$$F_{30} \cdot x_{7,A} + F_{24} \cdot x_{8,A} + F_{34} \cdot x_{3,A} + F_{22} + F_{28} \cdot x_{13,A}$$
$$+ F_{38} \cdot x_{14,A} - F_{31} \cdot x_{9,A} - F_{25} \cdot x_{9,A} = 0$$

$$F_{30} \cdot x_{7,B} + F_{24} \cdot x_{8,B} + F_{34} \cdot x_{3,B} + F_{28} \cdot x_{13,B}$$
$$+ F_{38} \cdot x_{14,A} - F_{31} \cdot x_{9,B} - F_{25} \cdot x_{9,B} = 0$$

$$F_{31} \cdot x_{9,A} + F_{25} \cdot x_{10,A} - F_{10} \cdot x_{4,A} = 0$$

$$F_{31} \cdot x_{9,B} + F_{25} \cdot x_{10,B} - F_{10} \cdot x_{4,B} = 0$$

$$F_{32} + F_{20} + F_{26} + F_{36} - F_{29} - F_{23} = 0$$

$$F_{32} \cdot x_{3,A} + F_{20} + F_{26} \cdot x_{13,A} + F_{36} \cdot x_{14,A} - F_{23} \cdot x_{5,A} = 0$$

$$F_{32} \cdot x_{3,B} + F_{26} \cdot x_{13,B} + F_{36} \cdot x_{14,B} - F_{23} \cdot x_{5,B} = 0$$

$$y2 - 3 \cdot y3 = 0$$

$$F_9 \cdot x_{1,A} - F_9 \cdot x_{3,A} + y1 \cdot (\nu_{1,A} \cdot a_1 + \nu_{2,A} \cdot b_1 + \nu_{3,A} \cdot c_1 + \nu_{4,A} \cdot d_1) = 0$$

$$F_9 \cdot x_{1,B} - F_9 \cdot x_{3,B} + y1 \cdot (\nu_{1,B} \cdot a_1 + \nu_{2,B} \cdot b_1 + \nu_{3,B} \cdot c_1 + \nu_{4,B} \cdot d_1) = 0$$

$$F_{23} \cdot x_{5,A} - F_{23} \cdot x_{6,A} + y3 \cdot (\nu_{1,A} \cdot a_2 + \nu_{2,A} \cdot b_2 + \nu_{3,A} \cdot c_2 + \nu_{4,A} \cdot d_2) = 0$$

$$F_{23} \cdot x_{5,B} - F_{23} \cdot x_{6,B} + y3 \cdot (\nu_{1,B} \cdot a_2 + \nu_{2,B} \cdot b_2 + \nu_{3,B} \cdot c_2 + \nu_{4,B} \cdot d_2) = 0$$

$$F_{24} \cdot x_{7,A} - F_{24} \cdot x_{8,A} + y3 \cdot (\nu_{1,A} \cdot a_3 + \nu_{2,A} \cdot b_3 + \nu_{3,A} \cdot c_3 + \nu_{4,A} \cdot d_3) = 0$$

$$F_{24} \cdot x_{7,B} - F_{24} \cdot x_{8,B} + y3 \cdot (\nu_{1,B} \cdot a_3 + \nu_{2,B} \cdot b_3 + \nu_{3,B} \cdot c_3 + \nu_{4,B} \cdot d_3) = 0$$

$$F_{25} \cdot x_{9,A} - F_{25} \cdot x_{10,A} + y3 \cdot (\nu_{1,A} \cdot a_4 + \nu_{2,A} \cdot b_4 + \nu_{3,A} \cdot c_4 + \nu_{4,A} \cdot d_4) = 0$$

$$F_{25} \cdot x_{9,B} - F_{25} \cdot x_{10,B} + y3 \cdot (\nu_{1,B} \cdot a_4 + \nu_{2,B} \cdot b_4 + \nu_{3,B} \cdot c_4 + \nu_{4,B} \cdot d_4) = 0$$

$$a_1 - k_1 \cdot z_{1,A} = 0$$

$$a_2 - k_1 \cdot z_{2,A} = 0$$

$$a_3 - k_1 \cdot z_{3,A} = 0$$

$$a_4 - k_1 \cdot z_{4,A} = 0$$

$$b_1 - k_2 \cdot z_{1,B} = 0$$

$$b_2 - k_2 \cdot z_{2,B} = 0$$

$$b_3 - k_2 \cdot z_{3,B} = 0$$

$$b_4 - k_2 \cdot z_{4,B} = 0$$

$$c_1 - k_3 \cdot z_{1,B} = 0$$

$$c_2 - k_3 \cdot z_{2,B} = 0$$

$$c_3 - k_3 \cdot z_{3,B} = 0$$

$$c_4 - k_3 \cdot z_{4,B} = 0$$

$$d_1 - k_4 \cdot z_{1,C} = 0$$

$$d_2 - k_4 \cdot z_{2,C} = 0$$

$$d_3 - k_4 \cdot z_{3,C} = 0$$

$$d_4 - k_4 \cdot z_{4,C} = 0$$

$$F_7 + F_8 - F_{14} - F_{19} - F_{35} = 0$$

$$F_7 \cdot x_{3,A} + F_8 \cdot x_{4,A} - F_{14} \cdot x_{11,A} - F_{19} \cdot x_{11,A} - F_{35} \cdot x_{11,A} = 0$$

$$F_7 \cdot x_{3,B} + F_8 \cdot x_{4,B} - F_{14} \cdot x_{11,B} - F_{19} \cdot x_{11,B} - F_{35} \cdot x_{11,B} = 0$$

$$F_{14} - F_{15} - F_{17} = 0$$

$$x_{12,B} \cdot (x_{11,B} + x_{11,C}) - x_{11,B} = 0$$

$$F_{17} - F_{14} \cdot x_{11,A} = 0$$

$$F_{15} - F_{18} - F_{16} = 0$$

$$F_{18} - x_{12,B} \cdot F_{15} = 0$$

$$F_{11} + F_{12} - F_{19} - F_{17} + F_{13} = 0$$

$$F_{39} + F_{40} - F_{35} - F_{16} + F_{41} = 0$$

$$F_{11} \cdot x_{13,A} + F_{12} \cdot x_{13,A} - F_{19} \cdot x_{11,A} - F_{17} \cdot x_{11,A} + F_{13} \cdot x_{13,A} = 0$$

$$F_{11} \cdot x_{13,B} + F_{12} \cdot x_{13,B} - F_{19} \cdot x_{11,B} + F_{13} \cdot x_{13,B} = 0$$

$$F_{39} \cdot x_{14,A} + F_{40} \cdot x_{14,A} - F_{35} \cdot x_{11,A} + F_{41} \cdot x_{14,A} = 0$$

$$F_{39} \cdot x_{14,C} + F_{40} \cdot x_{14,C} - F_{35} \cdot x_{11,C} - F_{16} + F_{41} \cdot x_{14,C} = 0$$

$$x_{1,A} + x_{1,B} + x_{1,C} - 1 = 0$$

$$x_{2,A} + x_{2,B} + x_{2,C} - 1 = 0$$

$$x_{3,A} + x_{3,B} + x_{3,C} - 1 = 0$$

$$x_{4,A} + x_{4,B} + x_{4,C} - 1 = 0$$

$$x_{5,A} + x_{5,B} + x_{5,C} - 1 = 0$$

$$x_{6,A} + x_{6,B} + x_{6,C} - 1 = 0$$

$$x_{7,A} + x_{7,B} + x_{7,C} - 1 = 0$$

$$x_{8,A} + x_{8,B} + x_{8,C} - 1 = 0$$

$$x_{9,A} + x_{9,B} + x_{9,C} - 1 = 0$$

$$x_{10,A} + x_{10,B} + x_{10,C} - 1 = 0$$

$$x_{11,A} + x_{11,B} + x_{11,C} - 1 = 0$$

$$x_{12,B} + x_{12,C} - 1 = 0$$

$$x_{13,A} + x_{13,B} + x_{13,C} - 1 = 0$$

$$x_{14,A} + x_{14,B} + x_{14,C} - 1 = 0$$

$$v_1 - p_1 \cdot x_{3,A} + p_2 \cdot x_{3,B} + p_3 \cdot x_{3,C} = 0$$

$$v_2 - p_1 \cdot x_{6,A} + p_2 \cdot x_{6,B} + p_3 \cdot x_{6,C} = 0$$

$$v_3 - p_1 \cdot x_{8,A} + p_2 \cdot x_{8,B} + p_3 \cdot x_{8,C} = 0$$

$$v_4 - p_1 \cdot x_{10,A} + p_2 \cdot x_{10,B} + p_3 \cdot x_{10,C} = 0$$

$$z_{1,A} \cdot v_1 - x_{3,A} = 0$$

$$z_{2,A} \cdot v_2 - x_{6,A} = 0$$

$$z_{3,A} \cdot v_3 - x_{8,A} = 0$$

$$z_{4,A} \cdot v_4 - x_{10,A} = 0$$

$$z_{1,B} \cdot v_1 - x_{3,B} = 0$$

$$z_{2,B} \cdot v_2 - x_{6,B} = 0$$

$$z_{3,B} \cdot v_3 - x_{8,B} = 0$$

$$z_{4,B} \cdot v_4 - x_{10,B} = 0$$

$$z_{1,C} \cdot v_1 - x_{3,C} = 0$$
$$z_{2,C} \cdot v_2 - x_{6,C} = 0$$
$$z_{3,C} \cdot v_3 - x_{8,C} = 0$$
$$z_{4,C} \cdot v_4 - x_{10,C} = 0$$
$$F_{18} - 50 \geq 0$$
$$r_1 - s_4 - F_{14} \cdot (s_5 + s_6 \cdot x_{11,A} + s_7 \cdot x_{11,B}) = 0$$
$$r_2 - s_8 - F_{15} \cdot (s_9 \cdot x_{12,B} + s_{10} \cdot x_{12,B}^2) = 0$$

10.3.2 Data

– stoichiometric coefficients: $\nu_{rp,m}$

	A	B	C
1	-1	1	0
2	1	-1	0
3	0	-1	1
4	0	1	-1

– kinetic constants: k_i

	k_i
1	0.4
2	0.4
3	0.4
4	0.4

– cost parameters

Parameter	Value	Parameter	Value
s_1	10,208.344	s_6	418.420
s_2	5,623.912	s_7	-152.456
s_3	0.4	s_8	54,966.389
s_4	41,357.32	s_9	748.609
s_5	432.709	s_{10}	-588.673

– molecular volumes p_i

	p_i
1	0.0885
2	0.0867
3	0.0862

- Upper and lower bounds

Variable			Lower	Upper
F_i		$i = 2, 3, 20, 21, 22$	0.0	100
F_i		$i = 4, \ldots, 19, 20, \ldots, 41$	0.0	200
$x_{i,m}$	$m = A, B, C$	$i = 1, \ldots, 14$	0.0	1.0
$z_{i,A}$		$i = 1, \ldots, 4$	0.0	12.0
$z_{i,B}$		$i = 1, \ldots, 4$	0.0	12.0
$z_{i,C}$		$i = 1, \ldots, 4$	0.0	6.0
$y1$			0.2	500
$y3$			0.001	4.0

10.3.3 Problem Statistics

Number of Variables	120
Number of Linear Constraints	52
Number of Nonlinear Constraints	43

10.3.4 Best Known Solution

- Objective value=85,426.965

- Flowrates

Var.	Level	Var.	Level	Var.	Level	Var.	Level
F_1	22.00	F_2	0.0	F_3	22.00	F_4	0.0
F_5	50.451	F_6	0.0	F_7	50.451	F_8	0.0
F_9	50.451	F_{10}	50.451	F_{11}	0.0	F_{12}	27.094
F_{13}	10.391	F_{14}	50.451	F_{15}	12.965	F_{16}	2.965
F_{17}	37.486	F_{18}	10.0	F_{19}	0.0	F_{20}	22.00
F_{21}	0.0	F_{22}	0.0	F_{23}	50.451	F_{24}	50.451
F_{25}	50.451	F_{26}	27.094	F_{27}	0.0	F_{28}	0.0
F_{29}	0.0	F_{30}	0.0	F_{31}	0.0	F_{32}	0.0
F_{33}	0.0	F_{34}	0.0	F_{35}	0.0	F_{36}	1.357
F_{37}	0.0	F_{38}	0.0	F_{39}	0.0	F_{40}	1.357
F_{41}	1.609						

- Concentrations and molar fractions

Var.	Level	Var.	Level	Var.	Level	Var.	Level
$z_{1,A}$	8.443	$z_{2,A}$	10.225	$z_{3,A}$	9.551	$z_{4,A}$	8.964
$z_{1,B}$	2.252	$z_{2,B}$	0.759	$z_{3,B}$	1.368	$z_{4,B}$	1.855
$z_{1,C}$	0.668	$z_{2,C}$	0.340	$z_{3,C}$	0.420	$z_{4,C}$	0.531
$x_{1,A}$	0.790	$x_{2,A}$	0.973	$x_{3,A}$	0.743	$x_{4,A}$	0.790
$x_{1,B}$	0.163	$x_{2,B}$	0.00	$x_{3,B}$	0.198	$x_{4,B}$	0.163
$x_{1,C}$	0.047	$x_{2,C}$	0.027	$x_{3,C}$	0.059	$x_{4,C}$	0.047
$x_{5,A}$	0.973	$x_{6,A}$	0.903	$x_{7,A}$	0.903	$x_{8,A}$	0.842
$x_{5,B}$	0.00	$x_{6,B}$	0.067	$x_{7,B}$	0.067	$x_{8,B}$	0.121
$x_{5,C}$	0.027	$x_{6,C}$	0.030	$x_{7,C}$	0.030	$x_{8,C}$	0.037
$x_{9,A}$	0.842	$x_{10,A}$	0.790	$x_{11,A}$	0.743	$x_{12,A}$	0.00
$x_{9,B}$	0.121	$x_{10,B}$	0.163	$x_{11,B}$	0.198	$x_{12,B}$	0.771
$x_{9,C}$	0.037	$x_{10,C}$	0.047	$x_{11,C}$	0.059	$x_{12,C}$	0.229
$x_{13,A}$	1.00	$x_{13,B}$	0.00	$x_{13,C}$	0.00	$x_{14,A}$	0.00
$x_{14,B}$	0.00	$x_{14,C}$	1.00				

- Reaction rates, volumes and molecular volumes

Var.	Level	Var.	Level	Var.	Level	Var.	Level
a_1	3.377	a_2	4.090	a_3	3.820	a_4	3.586
b_1	0.901	b_2	0.304	b_3	0.547	b_4	0.749
c_1	0.901	c_2	0.304	c_3	0.547	c_4	0.749
d_1	0.267	d_2	0.136	d_3	0.168	d_4	0.213
v_1	0.088	v_2	0.088	v_3	0.088	v_4	0.088
r_1	77,348.093	r_2	57,912.119	y1	0.952	y2	2.803
y3	0.934						

Chapter 11

Mechanical Design test problems

In this chapter, a few test problems that arise in certain applications of mechanical design are presented .

11.1   Test Problem 1 : Pressure Vessel Design

This test problem is taken from [136] and involves a horizontal pressure vessel for storage of liquid butane mounted on two equidistant saddle supports that is to be sized so as to minimize the cost of manufacture using the ASME code stipulations as constraints. The vessel has a hollow cylindrical shell of 50 feet length and 5 feet diameter, and its thickness is assumed to be small relative to other dimensions. Design variables are the thicknesses for head T_H , shell T_S , saddle ring T_{SR} , vacuum ring T_{VR} , and wear plate T_{WP} . They are formed from plates available only on thickness increments of (1/16) inch. The cost of manufacture varies linearly with T_S , T_H , and T_{WP} , and quadratically with respect to T_{SR} and T_{VR} .

11.1.1 Problem Formulation

$$MIN \quad f = 19380 + 32154T_S + 13189T_H + 2376T_{WP} + 5329\left(T_{VR}^2 + 2T_{SR}^2\right)$$

Subject to

$$\frac{P_i R}{2T_S} + \frac{Q L_A}{\pi R^2 T_S}\left[1 - \frac{1 - \frac{L_A}{12L} + \frac{R^2 - H^2}{24 L_A L}}{1 + \frac{H}{9L}}\right] - \sigma_A \leq 0$$

$$\frac{P_i R}{2T_S} + \frac{3Q L}{\pi R^2 \left\{T_S + \frac{A_{VR}}{L_{RR}}\right\}}\left[\frac{1 + 2\frac{R^2 - H^2}{(12L)^2}}{1 + \frac{H}{9L}} - \frac{L_A}{3L}\right] - E_F \sigma_A \leq 0$$

$$\frac{P_i R}{2T_S} - \frac{Q L_A}{0.603 R^2 T_S}\left[1 - \frac{1 - \frac{L_A}{12L} + \frac{R^2 - H^2}{24 L_A L}}{1 + \frac{H}{9L}}\right] - \frac{E_Y T_S}{29R}\left\{2 - \frac{200 T_S}{3R}\right\} \leq 0$$

$$\frac{P_i R}{2T_S} - \frac{3Q L}{\pi R^2 \left\{T_S + \frac{A_{VR}}{L_{RR}}\right\}}\left[\frac{1 + 2\frac{R^2 - H^2}{(12L)^2}}{1 + \frac{H}{9L}} - \frac{L_A}{3L}\right] - \frac{E_Y T_S}{29R}\left\{2 - \frac{200 T_S}{3R}\right\} \leq 0$$

$$\frac{P_i R}{2T_S} - \frac{Q L_A}{0.603 R^2 T_S}\left[1 - \frac{1 - \frac{L_A}{12L} + \frac{R^2 - H^2}{24 L_A L}}{1 + \frac{H}{9L}}\right] - 0.5\sigma_y \leq 0$$

$$\frac{P_i R}{2T_S} - \frac{3Q L}{\pi R^2 \left\{T_S + \frac{A_{VR}}{L_{RR}}\right\}}\left[\frac{1 + 2\frac{R^2 - H^2}{(12L)^2}}{1 + \frac{H}{9L}} - \frac{L_A}{3L}\right] - 0.5\sigma_y \leq 0$$

$$\frac{K_7 Q}{(T_S + T_{WP})(B + 1.56\sqrt{RT_S})} - 0.5\sigma_Y \leq 0$$

$$\frac{K_3 Q}{R(T_S + T_{WP})}\left[\frac{12L - 2L_A}{12L + \frac{4}{3}H}\right] - 0.8\sigma_A \leq 0$$

$$\frac{1}{T_H} - 0.8\sigma_A \leq 0$$

$$\frac{K_{10} Q R L_C}{I_{SR}} - \frac{K_9 Q}{A_{SR}} - 0.5\sigma_{yR} \leq 0$$

$$\frac{K_{10} Q R L_D}{I_{SR}} + \frac{K_9 Q}{A_{SR}} - 0.5\sigma_{yR} \leq 0$$

$$\frac{3K_{11} Q}{R T_{WEB}} - \frac{2}{3}\sigma_{AS} \leq 0$$

$$\delta_{ms} - 0.024 L \leq 0$$

$$4 + \frac{B}{2} + 0.78\sqrt{RT_S} - L_A \leq 0$$

$$T_S - T_H - 0.125 \leq 0$$

$$T_S - T_{WP} \leq 0$$

$$T_S - \frac{1}{5}R \leq 0$$

$$T_H - \frac{1}{5}R \leq 0$$

$$I_{VRM} - I_{VR} \leq 0$$

$$I_{VRM} - I_{SR} \leq 0$$

$$P_E - \frac{E_Y}{16}\left\{\frac{T_H}{0.9DOH}\right\}^2 \leq 0$$

$$P_E - \frac{0.93E_Y L_{RR}}{DOS}\left\{\frac{T_S}{DOS}\right\}^{2.5} \leq 0$$

where the intermediate variables are defined by the formulae below :

$$DOS = D + 2T_S$$

$$DOH = D + 2T_H$$

$$L_{RR} = 0.5(12L) - L_A$$

$$W_{ves} = 1.1\rho_s\left[12\pi DL(T_S + T_{CA}) + 2.16D^2(T_H + T_{CA})\right]$$

$$W_{liq} = \rho_b(3\pi D^2 L + 0.262D^3)$$

$$Q = (W_{ves} + W_{liq})/2$$

$$X_{ie} = \pi T_S(D + 2T_S)^{3/8}$$

$$W_u = 2Q/(12L + 4H/3)$$

$$\delta_{ms} = W_u(12L - 2L_A)^2\left\{5(12L - 2L_A)^2 - 24(2H/3 + L_A)^2\right\}/(384E_Y X_{ie})$$

$$A_1 = \left(T_{SR} + 1.56\sqrt{RT_S}\right)T_S$$

$$A_2 = 8T_{SR}^2$$

$$A_{SR} = A_1 + A_2$$

$$L_C = A_1 T_S/2 + A_2(T_S + 4T_{SR})/A_{SR}$$

$$L_D = T_S + 8_{SR} - L_C$$

$$H_1 = L_C - T_S/2$$

$$H_2 = T_S + 4T_{SR} - L_C$$

$$I_{SR} = A_1 H_1^2 + A_2 H_2^2 + A_1 T_S^2/12 + T_{SR}(8T_{SR})^3/12$$

$$A_{1V} = 1.56T_S\sqrt{RT_S}$$

$$A_{2V} = 8T_{VR}^2$$

$$A_{VR} = A_{1V} + A_{2V}$$

$$L_{CV} = A_{1V}T_S/2 + A_{2V}(T_S + 4T_{VR})/A_{VR}$$

$$L_{DV} = T_S + 8T_{VR} - L_{CV}$$

$$H_{1V} = L_{CV} - T_S/2$$

$$H_{2V} = T_S + 4T_{VR} - L_{CV}$$

$$I_{VR} = A_{1V}H_{1V}^2 + A_{2V}H_{2V}^2 + A_{1V}T_S^2/12 + T_{VR}(8T_{VR})^3/12$$

$$I_{VRM} = 0.12798D^3(T_S + A_{VR}/L_{RR})(T_S/DOS)^{1.5}$$

11.1.2   Data

The parameter values for this problem are :

$\sigma_A = 17500$	$E_F = 0.85$	$K_{11} = 0.204$
$\sigma_{AR} = 12700$	$E_Y = 29*10^6$	$L = 50$
$\sigma_{AS} = 12700$	$H = 30$	$L_A = 50$
$\sigma_y = 38000$	$K_3 = 0.319$	$P_e = 5$
$\sigma_{yR} = 36000$	$K_4 = 0.880$	$P_i = 50$
$\rho_S = 0.28$	$K_7 = 0.760$	$R = 60$
$\rho_b = 0.0216$	$K_9 = 0.340$	$T_{CA} = 0.125$
$D = 120$	$K_{10} = 0.053$	$T_{WEB} = 0.5$

11.1.3   Problem Statistics

The statistics for this Test Problem are given as follows :

$$No. of Continuous Variables = 17$$

$$No. of Linear Constraints = 20$$

$$No. of Nonlinear Constraints = 28$$

11.1.4 Best Known Solution

The best known solution for this Test Problem has an objective function value of 47,818. The values of the variables are shown below :

$$T_H = \frac{5}{16}$$
$$T_S = \frac{3}{16}$$
$$T_{SR} = \frac{9}{16}$$
$$T_{VR} = \frac{18}{16}$$
$$T_{WP} = \frac{5}{16}$$

11.2 Test Problem 2 : Selection of bolts in standard sizes

This test problem is taken from [136] and involves the selection of standard size bolts to fasten flange joints of a pressure vessel undergoing rapid pressure fluctuation. An even number of bolts are placed uniformly at diameter distance of 350mm around a cylindrical pressure vessel of 250mm internal diameter. Internal gage pressure fluctuates rapidly between 0 and 2.5MPA. The objective is to minimize the total cost which consists of the price of bolts and the labor cost for drilling holes and installing the bolts. Three discrete variables x_1 , x_2 , and x_3 are introduced to describe the bolt selection and one discrete variable x_4 to represent the number of bolts.

11.2.1 Problem Formulation

$$MIN \qquad f = 12dn + 19n$$

Subject to

$$g_1 \; : \; x_1 \le 1$$

$$g_1 \quad : \quad \frac{FK}{2nA} - \sigma \le 0$$

$$g_1 \quad : \quad \frac{C}{nd} - 10 \le 0$$

$$g_1 \quad : \quad 5 - \frac{C}{nd} \le 0$$

where

$$n = 2x_4$$

$$d = x_1 x_2 + (1 - x_1)x_3$$

$$A = 2.857 - 1.0692d + 0.6562d^2$$

$$K = 0.3333$$

$$F = 245400$$

$$\sigma = 69$$

$$C = 350\pi$$

11.2.2 Problem Statistics

The statistics for this Test Problem are given as follows :

$$\text{No. of Continuous Variables} = 7$$
$$\text{No. of Linear Constraints} = 2$$
$$\text{No. of Nonlinear Constraints} = 5$$

11.2.3 Best Known Solution

The best known solution for this Test Problem has an objective function value of 360. The values of the nonzero variables are shown below :

$$x_1 = 1$$
$$x_2 = 3$$
$$x_3 = 7$$
$$x_4 = 4$$

11.3 Test Problem 3 : Weight Minimization of a Speed Reducer

This test problem is taken from [48], and involves the design of a speed reducer for small aircraft engine. The resulting optimization problem has the following form :

11.3.1 Problem Formulation

$$MIN \quad f(x) = 0.7854x_1x_2^2(3.3333x_3^2 + 14.9334x_3 - 43.0934)$$
$$-1.508x_1(x_6^2 + x_7^2) + 7.477(x_6^3 + x_7^3) +$$
$$0.7854(x_4x_6^2 + x_5x_7^2)$$

Subject to

$$x_1x_2^2x_3 \geq 27$$
$$x_1x_2^2x_3^2 \geq 397.5$$
$$x_2x_6^4x_3x_4^{-3} \geq 1.93$$
$$x_2x_7^4x_3x_5^{-3} \geq 1.93$$
$$A_1B_1-1 \leq 1100$$

where

$$A_1 = [(745x_4x_2^{-1}x_3^{-1})^2 + 16.9110^6]^{0.5}$$
$$B_1 = 0.1x_6^3$$
$$A_2B_2^{-1} \leq 850$$
$$A_2 = [(745x_5x_2^{-1}x_3^{-1})^2 + 157.510^6]^{0.5}$$
$$B_2 = 0.1x_7^3$$
$$x_2x_3 \leq 40$$
$$x_1x_2^{-1} \geq 5$$
$$x_1x_2^{-1} \leq 12$$

$$1.5x_6 - x_4 \leq -1.9$$
$$1.5x_7 - x_5 \leq -1.9$$
$$2.6 \leq x_1 \leq 3.6$$
$$0.7 \leq x_2 \leq 0.8$$
$$17 \leq x_3 \leq 28$$
$$7.3 \leq x_4 \leq 8.3$$
$$7.3 \leq x_5 \leq 8.3$$
$$2.9 \leq x_6 \leq 3.9$$
$$5 \leq x_7 \leq 5.5$$

11.3.2 Best Known Solution

The best known solution for this Test Problem has an objective function value of 2994.47. The values of the nonzero variables are shown below :

$$x_1 = 3.5$$
$$x_2 = 0.7$$
$$x_3 = 17$$
$$x_4 = 7.3$$
$$x_5 = 7.71$$
$$x_6 = 3.35$$
$$x_7 = 5.287$$

Chapter 12

VLSI Design test problems

Many aspects of physical chip design can be formulated as global optimization problems (in most cases quadratic problems with nonlinear constraints). For instance, the compaction problem can be stated as a global minimization problem of a nonconvex function :

Determine the minimum chip area subject to linear or nonlinear constraints

The constraints result from geometric design rules, from distance and connectivity requirements between various components of the circuit, and from user specified constraints.

The key feature of these problems is their large size. Next, a sample of typical problem formulations is provided. Additional information can be found in [138], [210], and [238]. For a more complete survey of optimization techniques utilized in integrated circuit design see [38].

12.1 Test Problem 1

Placement algorithms for integrated circuit layout which are optimal are known to be NP-complete. As a result many heuristics and different modeling techniques have been proposed. This example is taken from [36]. Using a quadratic metric for placement the following nonconvex quadratically constrained optimization problem arises :

12.1.1 Problem Formulation

$$MIN \quad f(x,y) = x^T B x + y^T B y$$

$$s.t. \quad e^T x = e^T y = 0$$
$$x^T x = y^T y = 1$$
$$x^T y = 0$$
$$x, y \in R^N$$

where the matrix B is positive semi-definite and its entries are given by :

$$B_{ij} = \delta_{ij} \sum_{k=1}^{N} C_{ik} - C_{ij}$$

where δ_{ij} is the Kronecker's delta, and C_{ij} is a symmetric matrix representing the connectivity between devices i and j. The variables x_i, y_i are the coordinates of some reference point on device i, and N is the total number of components.

12.2 Test Problem 2

The layout problem requires a combination of space and communication costs to be minimized. The special problem of planar rectangular spaces occurs, for example, in floor plans for electronic planar packages and for buildings. This example corresponds to the rectangular dualization problem and is taken from [138].

12.2.1   Problem Formulation

$$MIN \quad x_a y_a + x_b y_b + x_c y_c + x_d y_d + x_e y_e + x_f y_f$$

$$
\begin{aligned}
s.t. \quad -x_d + x_e &= 0 \\
-x_a + x_b - x_d + x_e &= 0 \\
-x_a + x_c - x_d + x_f &= 0 \\
-y_c + y_f &= 0 \\
-y_a - y_b - y_c + y_d + y_e + y_f &= 0 \\
x - a &\geq 5 \\
x - b &\geq 5 \\
x - c &\geq 2 \\
x - d &\geq 4 \\
x - e &\geq 4 \\
x - f &\geq 5 \\
y - a &\geq 5 \\
y - b &\geq 2 \\
y - c &\geq 5 \\
y - d &\geq 4 \\
y - e &\geq 5 \\
y - f &\geq 5 \\
x_b - x_c &\geq 1 \\
-y_d + y_a &\geq 1 \\
x_a y_a &\geq 30 \\
x_b y_b &\geq 20 \\
x_c y_c &\geq 20 \\
x_d y_d &\geq 25 \\
x_e y_e &\geq 15 \\
x_f y_f &\geq 20
\end{aligned}
$$

12.3 Test Problem 3

This test problem is taken from [194] and represents the linear placement problem. The linear placement problem consists of determining a vector $x \in R^n$ where each component x_i is the position on the line of the module i, such that the following objective function :

$$\sum_i \sum_j c_{ij}(x_i - x_j)^2$$

is minimized under the restrictions that no two modules can occupy the same position and the values of x are constrained to be the legal positions, that is the coordinates of the centers of the available slots.

12.3.1 Problem Formulation

$$min \quad x^T B x$$

$$s.t. \quad \sum_{i=1}^{n} x_i = \sum_{i=1}^{n} l_i$$

$$\sum_{i=1}^{n} x_i^2 = \sum_{i=1}^{n} l_i^2$$

$$\cdots\cdots$$

$$\sum_{i=1}^{n} x_i^n = \sum_{i=1}^{n} l_i^n$$

where l_i, $i = 1, ..., n$ are the legal positions, and $B = D - C$ is a symmetric matrix, $C = C_{ij}$, and D is a diagonal matrix with $d_{ii} = \sum_{j=1}^{n} C_{ij}$, $i = 1, ..., n$.

Bibliography

[1] Abrham, J. and R.N. Buie. A note on nonconcave continuous programming. *Zeit. fur Oper. Research*, 3:107–114, 1975.

[2] Achenie, L.K.E. and L.T. Biegler. Algorithmic synthesis of chemical reactor networks using mathematical programming. *Ind. Eng. Chem. Fundam.*, 25:621–627, 1986.

[3] Aggarwal, A. and C.A. Floudas. A decomposition approach for global optimum search in qp, nlp and minlp problems. *To appear, Annals of Oper. Res.*, 1990.

[4] Aggarwal, A. and C.A. Floudas. Synthesis of general distillation sequences - nonsharp separation. *Computers and Chemical Engineering*, 14, 1990.

[5] Al-Khayyal, F.A. and J.E. Falk. Jointly constrained biconvex programming. *Math. of Oper. Res.*, 8(2):273 – 286, 1983.

[6] Al-Khayyal, F.A., R. Horst and P.M. Pardalos. Global optimization of concave functions subject to quadratic constraints : An application in nonlinear bilevel programming. *To appear, Annals of Oper. Res.*, 1990.

[7] Altman, M. Bilinear programming. *Bull. Acad. Polon. Sci. Ser. Sci. Math. Astronom. Phys.*, 16:741–745, 1968.

[8] Aluffi-Pentini, F., V. Parisi and F. Zirilli. Global optimization and stochastic differential equations. *JOTA*, 47(1):1–17, 1985.

[9] Aneja, Y.P., V. Aggarwal and K.P.K. Nair. On a class of quadratic programs. *Europ. Journ. of Oper. Res.*, 18:62–72, 1984.

[10] Arrow, K.J. and A.D. Enthoven. Quasiconcave programming. *Econometrica*, 29:779–800, 1961.

[11] Arthur, J.L. and J.O. Frendewey. Generating travelling salesman problems with known optimal tours. *J. Opl. Soc.*, 39(2):153–159, 1988.

[12] Babaev, R.D. The construction of test problems in integer-value programming with binary unknowns. *USSR Comput. Maths. Math. Phys.*, 25(1):98–102, 1985.

[13] Babaev, R.D. Generation of test problems on covering and partition. *Zh. Vychisl. Mat. i Mat. Fiz.*, 27(9):1349–1359, 1987.

[14] Baker, T.E. and L.S. Lasdon. Successive linear programming at exxon. *Manag. Sci.*, 31:264, 1985.

[15] Balas, E. Nonconvex quadratic programming via generalized polars. *SIAM J. Appl. Math.*, 28:335–349, 1975.

[16] Balinski, M.L. An algorithm for finding all vertices of polyhedral sets. *SIAM J.*, 9(1):72–89, 1961.

[17] Bansal, P.P. and S.E. Jacobsen. An algorithm for optimizing network flow capacity under economies of scale. *JOTA*, 15(5):565–586, 1975.

[18] Bansal, P.P. and S.E. Jacobsen. Characterization of local solutions for a class of nonconvex problems. *JOTA*, 15(5):549–564, 1975.

[19] Bard, J.F. and J.E. Falk. A separable programming approach to the linear complementarity problem. *Comput. and Oper. Res.*, 9(2):153–159, 1982.

[20] Barr, R.S., F. Glover and D. Klingman. A new optimization method for large scale fixed charge transportation problems. *Oper. Res.*, 29(3):448–463, 1981.

[21] Bartels, R.H. and N. Mahdavi-Amiri. On generating test problems for nonlinear programming algorithms. *SIAM J. Sci. Stat. Comput.*, 7(3):769–798, 1986.

[22] Basso, P. Iterative methods for the localization of the global maximum. *SIAM J. Numer. Anal.*, 19(4):781–792, 1982.

[23] Basso, P. Optimal search for the global maximum of functions with bounded seminorm. *SIAM J. Numer. Anal.*, 22(5):888–903, 1985.

[24] Beale, E.M.L. Branch and bound methods for numerical optimization of nonconvex functions. In *Compstat 1980 Proc. Fourth Symp. Comput. Stat., Edinburgh, 1980*, pages 11–20, Physica, Vienna, 1980.

[25] Beale, E.M.L. and J.J.H. Forrest. Global optimization using special ordered sets. *Math. Progr.*, 10:52–69, 1976.

[26] Beale, E.M.L. and J.J.H. Forrest. Global optimization as an extension of integer programming. In Dixon, L.C.W. and G.P.Szego, editors, *Towards Global Optimisation 2*, pages 131–149. North Holland Publ. Comp., 1978.

[27] Beale, E.M.L. and J.A. Tomlin. Special facilities in a general mathematical programming system for nonconvex problems using ordered sets of variables. In Lawrence, J., editor, *The fifth international conference on Oper. Res.*, pages 447–454, London, 1970. Tavistock Publications.

[28] Benacer, R. and D.T. Pham. Global maximization of a nondefinite quadratic function over a convex polyhedron. In J.-Hiriart-Urruty, editor, *FERMAT Days 1985: Mathematics for optimization*, pages 65–76. Elsevier Sci. Publishers, 1985.

[29] Benders, J.F. Partitioning procedures for solving mixed-variables programming problems. *Numer. Math.*, 4:238–252, 1962.

[30] BenSaad, S. *An algorithm for a class of nonlinear nonconvex optimization problems.* PhD thesis, University of California, Los Angeles, 1989.

[31] BenSaad, S. and S.E. Jacobsen. A level set algorithm for a class of reverse convex programs. *Annals of Operations Research (to appear)*, 1990.

[32] Benson, H.P. Algorithms for parametric nonconvex programming. *JOTA*, 38(3):316–340, 1982.

[33] Benson, H.P. On the convergence of two branch and bound algorithms for non convex programming problems. *JOTA*, 36(1):129–134, 1982.

[34] Benson, H.P. A finite algorithm for concave minimization over a polyhedron. *Naval Res. Log. Quarterly*, 32:165–177, 1985.

[35] Berkcan, E. *Performance biased placement for integrated circuit layout design*. PhD thesis, Electr. Eng. Dept., University of Rochester, 1986.

[36] Blanks, J.P. Near optimal placement using a quadratic objective function. In *Proceedings of the 22nd Design Automation Conference*, pages 609–615, 1985.

[37] Bokhari, S.H. On the mapping problem. *IEEE Trans. Comp.*, 30(3):207–214.

[38] Brayton, R.K., G.D. Hachtel and A.L. Sangiovanni-Vincentelli. A survey of optimization techniques for integrated-circuit design. In *Proceedings of the IEEE*, volume 69, pages 1334–1362, 1981.

[39] Bulatov, V.P. The immersion method for the global minimization of functions on the convex polyhedra. *Symposium on Eng. Math. and Applications (Beijing, China)*, pages 335–338, 1988.

[40] Bullatov, V.P. and L.I. Kasinskaya. Some methods of concave minimization on a convex polyhedron and their applications (russian). Methods of optimization and their applications. *"Nauka" Sibirsk. Otdel, Novosibirsk*, pages 71–80, 1982.

[41] Burdet, C.A. Deux modeles de minimisation d'une fonction economique concave. *R.I.R.O.*, 1:79–84, 1970.

[42] Burdet, C.A. Generating all the faces of a polyhedron. *SIAM Jour.*, 26(1):72–89, 1974.

[43] Cabot, A.V. Variations on a cutting plane method for solving concave minimization problems with linear constraints. *Naval Res. Logist. Quart.*, 21:265–274, 1974.

[44] Cabot, A.V. and R.L. Francis. Solving nonconvex quadratic minimization problems by ranking the extreme points. *Oper. Res.*, 18:82–86, 1970.

[45] Candler, W. and R.J. Townsley. The maximization of a quadratic function of variables subject to linear inequalities. *Manag. Science*, 10(3):515–523, 1964.

[46] Chang, M. G. and F. Shepardson. *An integer programming test problem generator*, volume 199 of *Lecture Notes in Economics and Mathematical Systems*, pages 146–160. 1982.

[47] Charnes, A., W.M. Raike, J.D. Stutz and A.S. Walters. On generation of test problems for linear programming codes. *Communications ACM*, 17(10):583–586, 1974.

[48] Chew, S.H. and Q. Zheng. *Integral Global Optimization*, volume 298 of *Lecture Notes in Economics and Mathematical Systems*. Springer-Verlag, 1988.

[49] Chung, C.S., M.S. Hung and W.O. Rom. A hard knapsack problem. *Naval Res. Logistics*, 35:85–98, 1988.

[50] Ciesieski, M.J. and E. Kinnen. An analytic method for compacting routing area in integrated circuits. In *Proceeding of the 19th Design Automation Conference, Las Vegas NV*, pages 30–37.

[51] Colville, A.R. A comparative study of nonlinear programming codes. In Kuhn, H.W., editor, *Princeton Symosium on Mathematical Programming*. Princeton Univ. Press, 1970.

[52] Crowder, H.P., Dembo R.S. and Mulvey J.M. Reporting computational experiments in mathematical programming. *Math. Progr.*, 15:316–329, 1978.

[53] Crowder, H.P., Dembo R.S. and Mulvey J.M. On reporting computational experiments with mathematical software. *ACM Trans. on Math. Soft.*, 5:193–203, 1979.

[54] Csendes, T. Nonlinear parameter estimation by global optimization - efficiency and reliability. *Acta Cybernetica*, 8(4):361–370, 1988.

[55] Czochralska, I. Bilinear programming. *Zastosow. Mat.*, 17:495–514, 1982.

[56] Czochralska, I. The method of bilinear programming for nonconvex quadratic problems. *Zastosow . Mat.*, 17:515–525, 1982.

[57] Dembo, R.S. A set of geometric programming test problems and their solutions. *Math. Progr.*, 10:192–213, 1976.

[58] Diewert, W.E., Avriel M. and I. Zang. Nine kinds of quasiconcavity and concavity. *Journal of Econ. Theory*, 25:397–420, 1981.

[59] Dixon, L.C.W. and G.P. Szego, editors. *Towards global optimisation.* North-Holland, Amsterdam, 1975.

[60] Dixon, L.C.W. and G.P. Szego, editors. *Towards global optimisation 2.* North-Holland, Amsterdam, 1978.

[61] Dyer, M.E. The complexity of vertex enumeration methods. *Math. Oper. Res.*, 8:381–402, 1983.

[62] Dyer, M.E. and L.G. Proll. An algorithm for determining all extreme points of a convex polytope. *Math. Progr.*, 12:81–96, 1977.

[63] Erickson, R.E., C.L. Monma and A.F. Veinot. Send and split method for minimum concave cost network flows. *Math. of Oper. Research*, 12(4):634–664, 1987.

[64] Evtushenko, G.Y. *Numerical Optimization Techniques, (Translation series in Math. and Engin.).* Optimization Software Inc., Publications Division, N.Y., 1985.

[65] Falk, J.E. Lagrange multipliers and nonconvex programs. *SIAM J. of Control and Optim.*, 7:534–545, 1969.

[66] Falk, J.E. A linear max-min problem. *Math. Progr.*, 5:169–188, 1973.

[67] Falk, J.E. and K.L. Hoffman. A successive underestimating method for concave minimization problems. *Math. of Oper. Res.*, 1:251–259, 1976.

[68] Falk, J.E. and R.M. Soland. An algorithm for separable nonconvex programming problems. *Manag. Sci.*, 15(9):550–569, 1969.

[69] Florian, M. Nonlinear cost network models in transportation analysis. *Math. Progr. Study*, 26:167–196, 1986.

[70] Florian, M. and P. Robillard. An implicit enumeration algorithm for the concave cost network flow problem. *Manag. Sci.*, 18(3):184–193, 1971.

[71] Florian, M., M. Rossin-Arthiat and D. de Verra. A property of minimum concave cost flows in capacited networks. *Canadian Journal of Oper. Res.*, 9:293–304, 1971.

[72] Floudas, C.A. Separation synthesis of multicomponent feed streams into multicomponent product streams. *AIChE J.*, 33(4):540–550, 1987.

[73] Floudas, C.A. and A. Aggarwal. A decomposition approach for global optimum search in the pooling problem. *To appear in ORSA J. On Computing*, 2(3), 1990.

[74] Floudas, C.A., A. Aggarwal and A.R. Ciric. Global optimum search for nonconvex nlp and minlp problems. *Computers and Chemical Engineering*, 13(10):1117–1132, 1989.

[75] Floudas, C.A. and S.H. Anastasiadis. Synthesis of distillation sequences with several multicomponent feed and product streams. *Chemical Engineering Science*, 43(9):2407–2419, 1988.

[76] Floudas, C.A. and A.R. Ciric. Strategies for overcoming uncertainties in heat exchanger network synthesis. *Computers and Chemical Engineering*, 13(10):1133–1152, 1989.

[77] Floudas, C.A., A.R. Ciric and I.E. Grossmann. Automatic synthesis of optimum heat exchanger network configurations. *AIChE J.*, 32(2):276–290, 1986.

[78] Floudas, C.A. and G.E. Paules. A mixed-integer nonlinear programming formulation for the synthesis of heat integrated distillation sequences. *Computers and Chemical Engineering*, 12(6):531–546, 1988.

[79] Forgo, F. Cutting plane methods for solving nonconvex quadratic problems. *Acta Cybernet.*, 1:171–192, 1972.

[80] Forgo, F. *Nonconvex Programming*. Akademiai Kiado, Budapest, 1988.

[81] Gallo, G., C. Sandi and C. Sodini. An algorithm for the min concave cost flow problem. *European J. of Oper. Res.*, 4:249–255, 1980.

[82] Gallo, G. and C. Sodini. Adjacent extreme flows and application to min concave cost flow problems. *Networks*, 9:95–122, 1979.

[83] Gallo, G. and A. Ulculu. Bilinear programming: an exact algorithm. *Math. Progr.*, 12:173–174, 1977.

[84] Galperin, E.A. The cubic algorithm. *J. of Math. Anal. and Appl.*, 112:635–640, 1985.

[85] Ge, R.P. and Y.F. Qin. A class of filled functions for finding global minimizers of a function of several variables. *JOTA*, 54(2):241–252, 1987.

[86] Gfrerer, H. Globally convergent decomposition methods for nonconvex optimization problems. *Computing*, 32:199–227, 1984.

[87] Glinsman, D.H. and J.B. Rosen. Constrained concave quadratic global minimization by integer programming. Technical Report TR-86-37, Dept. of Comp. Sci. University of Minnesota, 1986.

[88] Goldstein, A.A. and J.F. Price. On descent from local minima. *Mathematics of Computation*, 25:569–574, 1971.

[89] Griffith, R.E. and R.A. Stewart. A nonlinear programming technique for the optimization of continuous processesing systems. *Manag. Sci.*, 7:379, 1961.

[90] Groch, A., L.M. Vidigal and S.W. Director. A new global optimization method for electronic circuit design. *IEEE Trans. on Circiuts and Systems*, CAS-32(2), 1985.

[91] Guisewite, G.M. and P.M. Pardalos. Algorithms for the uncapacitated single-source minimum concave cost network flow problem. *To appear in Operational Research*, 1990.

[92] Guisewite, G.M. and P.M. Pardalos. Minimum concave-cost network flow problems : applications, complexity and algorithms. *To appear in Annals of Operations Research*, 1990.

[93] Hager, W.W., P.M. Pardalos, I.M. Roussos and H.D. Sahinoglou. Active constraints, indefinite quadratic test problems, and complexity. To appear in JOTA, 1990.

[94] Hanan, M. and J.M. Kurtzberg. A review of the placement and quadratic assignment problems. *SIAM Rev.*, 14:324–342, 1972.

[95] Hansen, E.R. Global optimization using interval analysis - the multidimensional case. *Numer. Math.*, 34(3):247–270, 1980.

[96] Hansen, P., B. Jaumard and S.-H. Lu. A framework for algorithms in globally optimal design. *ASME, J. of Mech. Trans. and Automation in Design*, 111:353–360, 1989.

[97] Hansen, P., B. Jaumard and S.-H. Lu. Global minimization of univariate functions by sequential polynomial approximation. *Internation J. of Computer Math.*, 28:183–193, 1989.

[98] Hansen, P., B. Jaumard and S.-H. Lu. An analytic approach for global optimization. To appear in Math. Progr., 1990.

[99] Haverly, C.A. Studies of the behabiour of recursion for the pooling problem. *ACM SIGMAP Bulletin*, 25:19, 1978.

[100] Hesse, R. A heuristic search procedure for estimating a global solution of nonconvex programming problems. *Op. Res.*, 21:1267–1280, 1973.

[101] Hiebert, K.L. An evaluation of mathematical software that solves systems of nonlinear equations. *ACM Trans. Math. Soft.*, 8:5–20, 1982.

[102] Hirsch, W.M. and A.J. Hoffman. Extreme varieties, concave functions, and the fixed charge problem. *Comm. Pure Appl. Math.*, XIV:355–369, 1961.

[103] Hoaglin, D.C., V.C. Klema and S.C. Peters. Exploratory data analysis in a study of the performance of nonlinear optimization routines. *ACM Trans. Math. Soft.*, 8:145–162, 1982.

[104] Hock, W. and K. Schittkowski. *Test examples for nonlinear programming codes*, volume 187 of *Lecture Notes in Economics and Mathematical Systems*. Springer Verlag, 1981.

[105] Hoffman, K.L. *A successive underestimating method for concave minimization problems*. PhD thesis, The George Washington University, 1975.

[106] Hoffman, K.L. A method for globally minimizing concave functions over convex sets. *Math. Progr.*, 20:22–32, 1981.

[107] Horn, F.J.M. and M.J. Tsai. The use of the adjoint variables in the development of improvement criteria for chemical reactors. *JOTA*, 1:131–145, 1967.

[108] Horst, R. An algorithm for nonconvex programming problems. *Math. Progr.*, 10:312–321, 1976.

[109] Horst, R. *A new branch and bound approach for concave minimization problems*, volume 41 of *Lecture Notes in Computer Sc.*, pages 330–337. 1976.

[110] Horst, R. A note on the convergence of an algorithm for nonconvex programming problems. *Math. Progr.*, 19:237–238, 1980.

[111] Horst, R. A note on functions, whose local minimum are global. *JOTA*, 36(3):457–463, 1982.

[112] Horst, R. On global minimization of concave functions: Introduction and survey. *Oper. Res. Spektrum*, 6:195–205, 1984.

[113] Horst, R. On the convexification of nonlinear programming problems: An applications oriented survey. *European J.of Oper. Res.*, 15:382–392, 1984.

[114] Horst, R. and H. Tuy. *Global Optimization: Deterministic Approaches*. Springer-Verlag, 1990.

[115] Hu, T.C. and M.T. Shing. A decomposition algorithm for circuit routing. *Math. Progr.Study*, 24:87–103, 1985.

[116] Jackson, R. Optimization of chemical reactors with respect to flow configuration. *JOTA*, 2:240–259, 1968.

[117] Jacobsen, S.E. Convergence of a tuy-type algorithm for concave minimization subject to linear inequality constraints. *Appl. Math. Optim.*, 7:1–9, 1981.

[118] Kalantari, B. and J.B. Rosen. Construction of large scale global minimum concave quadratic test problems. *JOTA*, 48:303–313, 1986.

[119] Katoh, N. and T. Ibaraki. A parametric characterization for minimization of a quasiconcave program. *Mathematical Programming*, 17:39–66, 1987.

[120] Kearfott, R.B. Some tests of generalized bisection. *ACM Trans. Math. Soft.*, 13(3):197–220, 1987.

[121] Kedem, G. and H.Watanabe. Optimization techniques for ic layout and compaction. In *Proceedings IEEE Intern. Conf. in Computer Design: VLSI in Computers*, pages 709–713, 1983.

[122] Kedem, G. and H. Watanabe. Optimization techniques for ic layout and compaction. In *Proceedings IEEE Intern. Conf. in Computer Design: VLSI in Computers*, pages 709–713, 1983.

[123] Kelley, J.E.Jr. The cutting plane method for solving convex programs. *J. Soc. Indust. Appl. Math.*, 8:703–712, 1960.

[124] Kokossis, A.C. and C.A. Floudas. Optimal synthesis of isothermal reactor-separator-recycle systems. *Submitted to Chemical Engineering Science*, 1990.

[125] Kokossis, A.C. and C.A. Floudas. Optimization of complex reactor networks-i. isothermal operation. *Chemical Engineering Science*, 45(3):595–614, 1990.

[126] Konno, H. A cutting plane algorithm for solving bilinear programs. *Math. Progr.*, 11:14–27, 1976.

[127] Konno, H. Maximization of a convex quadratic function subject to linear constraints. *Math. Progr.*, 11:117–127, 1976.

[128] Konno, H. Maximizing a convex quadratic function over a hypercube. *J. Oper. Res. Soc. Japan*, 23(2):171–189, 1980.

[129] Konno, H. An algorithm for solving bilinear knapsack problems. *J. Oper. Res. Soc. Japan*, 24(4):360–373, 1981.

[130] Kough, P.L. The indefinite quadratic programming problem. *Oper. Res.*, 27(3):516–533, 1979.

[131] Lasdon, L.S., A.D. Waren, S. Sarkar and F. Palacios-Gomez. Solving the pooling problem using generalized reduced gradient and successive linear programming algorithms. *ACM SIGMAP Bulletin*, 27:9, 1979.

[132] Lenard, M.L. and M. Minkoff. Randomly generated test problems for positive definite quadratic programming. *ACM Trans. Math. Soft.*, 10(1):86–96, 1984.

[133] Lengauer, T. On the solution of inequality systems relevant to ic-layout. In *Proceedings of the 8th Conference on Graphtheoretic Concepts in Computer Science (WG 82)*, pages 151–163, 1982.

[134] Levy, A.V. and S. Gomez. The tunneling algorithm for global minimization of functions. *SIAM J. Sci. Stat. Comput.*, 1985.

[135] Lin, L.S. and J. Allen. Minplex- a compactor that minimizes the bounding rectangle and individual rectangles in a layout. In *23rd Design Automation Conference*, pages 123–129, 1986.

[136] Loh, H.T. and P.Y. Papalambros. A sequential linearization approach for solving mixed discrete nonlinear design optimization problems. Technical Report UM-MEAM-89-08, University of Michigan, Department of Mechanical Engineering, 1989.

[137] Majthay, A. and A. Whinston. Quasiconcave minimization subject to linear constraints. *Discrete Math.*, 1:35–39, 1974.

[138] Maling, K., S.H. Mueller and W.R. Heller. On finding most optimal rectangular package plans. In *Proceeding of the 19th Design Automation Conference*, pages 663–670, 1982.

[139] Manas, M. An algorithm for a nonconvex programming problem. *Econom. Mathem. Obzor*, 4(2):202–212, 1968.

[140] Mancini, L. and G.P. McCormick. Bounding global minima. *Math. of Oper. Res.*, 1(1):50–53, 1976.

[141] Marple, D.P. and A.E. Gamal. *Optimal selection of transistor sizes in digital VLSI circuits. Advanced Research in VLSI.* Stanford Univ., 1987.

[142] Martos, B. Quadratic programming with a quasiconvex objective function. *Oper. Res.*, 19:87–97, 1971.

[143] Matson, M.D. Optimization of digital MOS VLSI circuits. In Fuchs, H., editor, *Chapel Hill Conference on VLSI.* Computer Sc. Press, 1985.

[144] May, J.H. and L.R. Smith. *The definition and generation of geometrically random linear constraint sets*, volume 199 of *Lecture Notes in Economics and Mathematical Systems.* 1982.

[145] McCormick, G.P. Attempts to calculate global solutions of problems that may have local minima. In F.A.Lootsma, editor, *Numerical Methods for Non-linear Optimization*, pages 209–221. Acad. Press, 1972.

[146] McCormick, G.P. Computability of global solutions to factorable nonconvex programs: Part i-convex underestimating problems. *Math. Progr.*, 10:147–175, 1976.

[147] McCormick, G.P. *Nonlinear Programming: Theory, Algorithms and Applications.* John Wiley and Sons, NY, 1983.

[148] Michelsen, M.L. The isothermal flash problem - part i. stability. *Fluid Phase Equilibria*, 9:1, 1982.

[149] Michelsen, M.L. The isothermal flash problem - part ii. phase-split calculation. *Fluid Phase Equilibria*, 9:1, 1982.

[150] Miele, A. and S. Gonzales. On the comparative evaluation of algorithms for mathematical programming problems. *Nonlinear Programming 3*, pages 337–359, 1978. New York.

[151] Mladineo, R.H. An algorithm for finding the global maximum of a multimodal, multivariate function. *Math. Progr.*, 34:188–200, 1986.

[152] Mockus, J. *Bayesian Approach to Global Optimization.* Kluwer Academic Publishers, 1989.

[153] Moore, R.E. *Methods and applications of interval analysis.* SIAM, Philadelphia, 1979.

[154] More, J.J. A collection of nonlinear model problems. Technical Report MCS-P60-0289, Argonne National Laboratory, 1989.

[155] More, J.J., B.S. Garbow and K.E. Hillstrom. Testing unconstrained optimization software. *ACM Trans. Math. Soft.*, 13:133–137, 1981.

[156] Mueller, R.K. A method for solving the indefinite quadratic programming problem. *Manag. Sci.*, 16:333–339, 1970.

[157] Murty, K.G. and S.N. Kabadi. Some np-complete problems in quadratic and nonlinear programming. *Math. Progr.*, 39:117–129, 1987.

[158] Palacios-Gomez, F., L.S. Lasdon and M. Engquist. Nonlinear optimization by successive linear programming. *Manag. Sci.*, 28:1106, 1982.

[159] Palubetskis, G.S. A generator of quadratic assignment test problems with a known optimal solution. *Zh. Vychisl. Mat. i Mat. Fiz.*, 28(11):1740–1743, 1988.

[160] Papalambros, P.Y. and D.J. Wilde. *Priciples of optimal design.* Cambridge University Press, 1988.

[161] Papoulias, S.A. and I.E. Grossmann. A structural optimization approach in process synthesis - ii. heat recovery networks. *Computers and Chemical Engineering*, 7:707–731, 1983.

[162] Pardalos, P.M. *Integer and Separable programming techniques for large-scale global optimization problems.* PhD thesis, Computer Science Department, Univ. Minnesota, Minneapolis, MN, 1985.

[163] Pardalos, P.M. On generating test problems for global optimization algorithms. Technical report, Computer Sci. Dept. The Pennsylvania State University, 1985.

[164] Pardalos, P.M. Generation of large-scale quadratic programs for use as global optimization test problems. *ACM Trans. on Math. Software*, 13(2):133–137, 1987.

[165] Pardalos, P.M. Quadratic programming defined on a convex hull of points. *BIT*, 28:323–328, 1988.

[166] Pardalos, P.M. Construction of test problems in quadratic bivalent programming. *ACM Trans. Math. Soft.*, 1990.

[167] Pardalos, P.M. and V. Crouse. A parallel algorithm for the quadratic assignment problem. In *Supecomputing'89 conference*, pages 351–360. ACM Press, 1989.

[168] Pardalos, P.M., J.H. Glick and J.B. Rosen. Global minimization of indefinite quadratic problems. *Computing*, 39:281–291, 1987.

[169] Pardalos, P.M. and A. Phillips. Global optimization approach to the maximum clique problem. *Int. J. Computer Math.*, 33:209–216, 1990.

[170] Pardalos, P.M., A.T. Phillips and J.B. Rosen. *Parallel Computing in Mathematical Programming*. Frontiers in Applied Mathematics, SIAM, 1990.

[171] Pardalos, P.M. and J.B. Rosen. Methods for global concave minimization: A bibliographic survey. *SIAM Rev.*, 28(3):367–379, 1986.

[172] Pardalos, P.M. and J.B. Rosen. *Constrained global optimization: Algorithms and applications*, volume 268 of *Lecture Notes in Computer Science*. Springer Verlag, 1987.

[173] Pardalos, P.M. and J.B. Rosen. Global optimization approach to the linear complementarity problem. *SIAM J. Scient. Stat. Computing*, 9(2):341–353, 1988.

[174] Pardalos, P.M. and J.B. Rosen, editors. *Special issue on Computational methods in global optimization*. Annals of Operations Research, 1990.

[175] Pardalos, P.M. and G. Schnitger. Checking local optimality in constrained quadratic programming is np-hard. *Oper. Res. Letters*, 7(1):33–35, 1988.

[176] Passy, U. Global solutions of mathematical programs with intrinsically concave functions. *JOTA*, 26(1):97–115, 1978.

[177] Paules, G.E. and C.A. Floudas. A new optimization approach for phase and chemical equilibrium problems, 1990.

[178] Pham, Dinh T. and El B. Souad. Algorithms for solving a class of nonconvex optimization problems: Methods of subgradients. In J.-Hiriart-Urruty, editor, *FERMAT Days 1985: Mathematics for optimization*, pages 249–271. Elsevier Sci. Publishers, 1985.

[179] Phillips, A.T. and J.B. Rosen. A parallel algorithm for constrained concave quadratic global minimization. *Math. Progr.*, 42:421–448, 1988.

[180] Pho, T.K. and L. Lapidus. Topics in computer-aided design : Part ii. synthesis of optimal heat exchanger networks by tree searching algorithms. *AIChE J.*, 19:1182–1189, 1973.

[181] Pijavski, S.A. An algorithm for finding the absolute extremum of a function. *USSR Comput.Math. and Math. Phys.*, pages 57–67, 1972.

[182] Pinter, J. Extended univariate algorithms for n-dimensional global optimization. *Computing*, 36:91–103, 1986.

[183] Pinter, J. Solving nonlinear equation systems via global partition and search: some experimental results. *Computing*, 43:309–323, 1990.

[184] Ratschek, H. *Computer methods for the range of functions*. Wiley, 1984.

[185] Ratschek, H. and J. Rokne. *New Computer Methods for Global Optimization*. Halsted Press, 1988.

[186] Reeves, G.R. Global minimization in non-convex all-quadratic programming. *Manag. Sci.*, 22:76–86, 1975.

[187] Ritter, K. Stationary points of quadratic maximum problems. *Z. Wahrscheinlich. verw. Geb.*, 4:149–158, 1965.

[188] Ritter, K. A method for solving maximum problems with a nonconcave quadratic objective function. *Z. Wahrscheinlihkeitstheorie Geb.*, 4:340–351, 1966.

[189] Rosen, J.B. Global minimization of a linearly constrained concave function by partition of feasible domain. *Math. Oper. Res.*, 8:215–230, 1983.

[190] Rosen, J.B. Computational solution of large-scale constrained global minimization problems. In P.T.Boggs, R.H. Byrd and R.B.Schnabel, editors, *Numerical Optimization*, pages 263–271, Phil., 1984. SIAM.

[191] Rosen, J.B. Performance of approximate algorithms for global minimization. *Math. Progr. Study*, 22:231–236, 1984.

[192] Rosen, J.B. and P.M. Pardalos. Global minimization of large-scale constrained concave quadratic problems by separable programming. *Math. Progr.*, 34:163–174, 1986.

[193] Rosen, J.B. and S. Suzuki. Construction of nonlinear programming test problems. *Communications ACM*, 8(2):113, 1965.

[194] Sanglovanni-Vincentelli, A. Automatic layout of integrated circuits. Technical report, Computer Science Dept., University of California, Berkeley, 1987.

[195] Schaible, S. On factored quadratic programs. *Zeitschrift fur Oper. Res.*, 17:179–181, 1973.

[196] Schittkowski, K. *More test examples for nonlinear programming codes*, volume 282 of *Lecture Notes in Economics and Mathematical Systems*. Springer Verlag, 1987.

[197] Sen, S. and H.D. Sherali. A branch and bound algorithm for extreme point mathematical programming problems. *Discrete Appl. Math.*, 11:265–280, 1985.

[198] Sen, S. and H.D. Sherali. On the convergence of cutting plane algorithms for a class of nonconvex mathematical programs. *Math. Progr.*, 31(1):42–56, 1985.

[199] Sherali, H. and S. Sen. A disjunctive cutting plane algorithm for the extreme point mathematical programming problems. *Opsearch (Theory)*, 22(2):83–94, 1985.

[200] Sherali, H.D. and C.M. Shetty. Deep cuts in disjunctive programming. *Naval Research Logis. Quart.*, 27:453–476, 1980.

[201] Sherali, H.D. and C.M. Shetty. A finitely convergent algorithm for bilinear programming problems using polar and disjunctive face cuts. *Math. Progr.*, 19:14–31, 1980.

[202] Sherali, H.D. and C.M. Shetty. *Optimization with disjunctive constraints.* Springer, N.Y., 1980.

[203] Sherali, H.D. and C.M. Shetty. A finitely convergent procedure for facial disjunctive programs. *Discrete Appl. Math.*, 4:135–148, 1982.

[204] Shor, N.Z. *Minimization methods for nondifferential functions.* Springer-Verlag, 1985.

[205] Shor, N.Z. One idea of getting global extremum in polynomial problems of mathematical programming. *Kibernetica, Kiev*, (5):102–106, 1987.

[206] Shubert, B. A sequential method seeking the global maximum of a function. *SIAM J. of Num. Anal.*, 9(3):379–388, 1972.

[207] Shvatal, V. Hard knapsack problems. *Oper. Res.*, 28:1402–1411, 1980.

[208] Soland, R.M. An algorithm for separable nonconvex programming problems ii: nonconvex constraints. *Manag. Sci.*, 17(11):759–773, 1971.

[209] Soland, R.M. Optimal facility location with concave costs. *Oper. Res.*, pages 373–382, 1974.

[210] Soukup, J. Circuit layout. In *Proceedings of the IEEE*, volume 69, pages 1281–1304, 1984.

[211] Stephanopoulos, G. and A.W. Westerberg. The use of hestenes' method of multipliers to resolve dual gaps in engineering system optimization. *JOTA*, 15:285–309, 1975.

[212] Sung, Y.Y. and J.B. Rosen. Global minimum test problem construction. *Math. Progr.*, 24:353–355, 1982.

[213] Swarup, K. Indefinite quadratic programming. *Cahiers du Centre d'Etudes de Res. Oper.*, 8:217–222, 1966.

[214] Swarup, K. Quadratic programming. *Cahiers du Centre d'Etudes de Res. Oper.*, 8:223–234, 1966.

[215] Taha, H.A. Concave minimization over a convex polyhedron. *Naval Res. Logist. Quart.*, 20(1):533–548, 1973.

[216] Thach, P.T. and H. Tuy. Parametric approach to a class of nonconvex global optimization problems. *Optimization*, 19:3–11, 1988.

[217] Thakur, L.S. Domain contraction in nonconvex programming: Minimizing a quadratic concave objective under linear constraints. To appear in Math. of Oper. Res., 1990.

[218] Thoai, N.V. and H. Tuy. Convergent algorithms for minimizing a concave function. *Math. of Oper. Res.*, 5(4):556–566, 1980.

[219] Thoai, N.V. and H. Tuy. Solving the linear complementarity problem through concave programming. *Zh. Vychisl. Mat. i. Mat. Fiz.*, 23(3):602–608, 1983.

[220] Torn, A. and A. Zilinskas. *Global Optimization*, volume 350 of *Lecture Notes in Computer Science*. Springer-Verlag, 1989.

[221] Trangenstein, J.A. Customized minimization techniques for phase equilibrium computations in reservoir simulation. *Chemical Engineering Science*, 42(12):2847, 1987.

[222] Tuy, H. Concave programming under linear constraints. *Dokl. Akad. Nauk. SSSR*, 159:32–35, 1964. Translated: Soviet Math. Dokl.5 1964 pp.1437–1440.

[223] Tuy, H. Global minimization of a difference of two convex functions. In *Selected Topics in Operations Research and Mathematical Economics, Lecture Notes Econ. Math. Syst. 226*, pages 98–118. 1984.

[224] Tuy, H. Concave minimization under linear constraints with special structure. *Optimization*, 16(3):335–352, 1985.

[225] Tuy, H. A general deterministic approach to global optimization via d.c. programming. In Hiriart-Urruty, J., editor, *FERMAT Days 1985: Mathematics for optimization*, pages 273–303. Elsevier Sci. Publishers, 1985.

[226] Tuy, H. and R. Horst. Convergence and restart in branch and bound algorithms for global optimization. application to concave minimization and d.c. programming. *Math. Progr.*, 41:161–183, 1988.

[227] Tuy, H. Thieu. T.V. and Thai. N.Q. A conical algorithm for globally minimizing a concave function over a closed convex set. *Math. of Oper. Res.*, 10(3):498–514, 1985.

[228] Vaish, H. and C.M. Shetty. The bilinear programming problem. *Naval Res. Logist. Quart.*, 23:303–309, 1976.

[229] Vaish, H. and C.M. Shetty. A cutting plane algorithm for the bilinear programming problem. *Naval Res. Logist. Quart.*, 24:83–94, 1977.

[230] Van Eijndhoven, J.T.J. Solving the linear complementarity problem in circuit simulation. *SIAM J. Control and Optim.*, 24(5):1050–1062, 1986.

[231] Vasil'ev, N.S. Search for a global minimum of a quasiconcave function. *USSR Comput. Maths. Math. Phys.*, 23(2):31–35, 1983.

[232] Vasil'ev, N.S. An active method of searching for the global minimum of a concave function. *USSR Comput. Maths. Math. Phys.*, 24(1):96–100, 1984.

[233] Vasil'ev, N.S. Minimum search in concave problems, using the sufficient condition for a global extremum. *USSR Comput. Maths. Math. Phys.*, 25(1):123–129, 1985.

[234] Veinott, A.F. Minimum concave cost solution of leontiev substitution models of multi-facility inventory systems. *Oper. Res.*, 14:486–507, 1969.

[235] Veinott, A.F. Existence and characterization of minima of concave functions on unbounded convex sets. *Math. Progr. Study*, 25:88–92, 1985.

[236] Verina, L.F. Solution of some nonconvex problems by reduction to nonconvex parametric programming. *Vestsi Akad. Navuk BSSR Ser. Fiz. Mat. Navuk*, 1:13–18, 1985.

[237] Walster, G.W., E.R. Hansen and S. Sengupta. Test results of a global optimization algorithm. *Numerical Optimization, SIAM (Boggs, P.T. et al editors)*, pages 272–287, 1984.

[238] Watanabe, H. *IC layout generation and compaction using mathematical programming*. PhD thesis, Computer Sc. Dept. Univ. of Rochester, 1984.

[239] Wehe, R.R. and A.W. Westerberg. An algorithmic procedure for the synthesis of distillation sequences with bypass. *Computers and Chemical Engineering*, 11:619, 1987.

[240] Westerberg, A.W. The synthesis of distillation based separation sequences. *Computers and Chemical Engineering*, 9:421, 1985.

[241] Westerberg, A.W. and J.V. Shah. Assuring a global minimum by the use of an upper bound on the lower (dual) bound. *Computers and Chemical Engineering*, 2:83–92, 1978.

[242] White, W.B., S.M. Johnson and G.B. Dantzig. Chemical equilibrium in complex mixtures. *J. Chem. Phys.*, 28(5):751, 1958.

[243] Wilde, D.J. *Optimum seeking methods*. Prentice Hall, Englewood Cliffs, N.J., 1964.

[244] Wingo, D.R. Globally minimizing polynomials without evaluating derivatives. *Intern. J. Computer Math.*, 17:287–294, 1985.

[245] Zaliznyak, N.F. and A.A. Ligun. Optimal strategies for seeking the global maximum of a function. *USSR Comp. Math. Math. Phys.*, 18(2):31–38, 1978.

[246] Zangwill, W.I. Minimum concave cost flows in certain networks. *Management Science*, 15(9):429–450, 1968.

[247] Zhirov, V.S. Search for the global extremum of a polynomial on a parallelepiped. *USSR Comput. Maths. Math. Phys.*, 25(1):105–116, 1985.

[248] Zilverberg, N.D. *Global minimization for large scale linear constrained systems.* PhD thesis, Computer Sc. Dept. Univ. of Minnesota, 1983.

[249] Zwart, P.B. Computational aspects of the use of cutting planes in global optimization. In *Proceeding of the 1971 annual conference of the ACM*, pages 457–465, 1971.

[250] Zwart, P.B. Nonlinear programming: Counterexamples to two global optimization algorithms. *Oper. Res.*, 21(6):1260–1266, 1973.

[251] Zwart, P.B. Global maximization of a convex function with linear inequality constraints. *Oper. Res.*, 22(3):602–609, 1974.

Vol. 408: M. Leeser, G. Brown (Eds.),Hardware Specification, Verification and Synthesis: Mathematical Aspects. Proceedings, 1989. VI, 402 pages. 1990.

Vol. 409: A. Buchmann, O. Günther, T. R. Smith, Y.-F. Wang (Eds.), Design and Implementation of Large Spatial Databases. Proceedings, 1989. IX, 364 pages. 1990.

Vol. 410: F. Pichler, R. Moreno-Diaz (Eds.), Computer Aided Systems Theory – EUROCAST '89. Proceedings, 1989. VII, 427 pages. 1990.

Vol. 411: M. Nagl (Ed.), Graph-Theoretic Concepts in Computer Science. Proceedings, 1989. VII, 374 pages. 1990.

Vol. 412: L. B. Almeida, C. J. Wellekens (Eds.), Neural Networks. Proceedings, 1990. IX, 276 pages. 1990.

Vol. 413: R. Lenz, Group Theoretical Methods in Image Processing. VIII, 139 pages. 1990.

Vol. 414: A.Kreczmar, A. Salwicki, M. Warpechowski, LOGLAN '88 – Report on the Programming Language. X, 133 pages. 1990.

Vol. 415: C. Choffrut, T. Lengauer (Eds.), STACS 90. Proceedings, 1990. VI, 312 pages. 1990.

Vol. 416: F. Bancilhon, C. Thanos, D. Tsichritzis (Eds.), Advances in Database Technology – EDBT '90. Proceedings, 1990. IX, 452 pages. 1990.

Vol. 417: P. Martin-Löf, G. Mints (Eds.), COLOG-88. International Conference on Computer Logic. Proceedings, 1988. VI, 338 pages. 1990.

Vol. 418: K.H. Bläsius, U. Hedtstück, C.-R. Rollinger (Eds.), Sorts and Types in Artificial Intelligence. Proceedings, 1989. VIII, 307 pages. 1990. (Subseries LNAI).

Vol. 419: K. Weichselberger, S. Pöhlmann, A Methodology for Uncertainty in Knowledge-Based Systems. VIII, 136 pages. 1990 (Subseries LNAI).

Vol. 420: Z. Michalewicz (Ed.), Statistical and Scientific Database Management, V SSDBM. Proceedings, 1990. V, 256 pages. 1990.

Vol. 421: T. Onodera, S. Kawai, A Formal Model of Visualization in Computer Graphics Systems. X, 100 pages. 1990.

Vol. 422: B. Nebel, Reasoning and Revision in Hybrid Representation Systems. XII, 270 pages. 1990 (Subseries LNAI).

Vol. 423: L. E. Deimel (Ed.), Software Engineering Education. Proceedings, 1990. VI, 164 pages. 1990.

Vol. 424: G. Rozenberg (Ed.), Advances in Petri Nets 1989. VI, 524 pages. 1990.

Vol. 425: C.H. Bergman, R.D. Maddux, D.L. Pigozzi (Eds.), Algebraic Logic and Universal Algebra in Computer Science. Proceedings, 1988. XI, 292 pages. 1990.

Vol. 426: N. Houbak, SIL – a Simulation Language. VII, 192 pages. 1990.

Vol. 427: O. Faugeras (Ed.), Computer Vision – ECCV 90. Proceedings, 1990. XII, 619 pages. 1990.

Vol. 428: D. Bjørner, C. A. R. Hoare, H. Langmaack (Eds.), VDM '90. VDM and Z – Formal Methods in Software Development. Proceedings, 1990. XVII, 580 pages. 1990.

Vol. 429: A. Miola (Ed.), Design and Implementation of Symbolic Computation Systems. Proceedings, 1990. XII, 284 pages. 1990.

Vol. 430: J. W. de Bakker, W.-P. de Roever, G. Rozenberg (Eds.), Stepwise Refinement of Distributed Systems. Models, Formalisms, Correctness. Proceedings, 1989. X, 808 pages. 1990.

Vol. 431: A. Arnold (Ed.), CAAP '90. Proceedings, 1990. VI, 285 pages. 1990.

Vol. 432: N. Jones (Ed.), ESOP '90. Proceedings, 1990. IX, 436 pages. 1990.

Vol. 433: W. Schröder-Preikschat, W. Zimmer (Eds.), Progress in Distributed Operating Systems and Distributed Systems Management. Proceedings, 1989. V, 206 pages. 1990.

Vol. 435: G. Brassard (Ed.), Advances in Cryptology – CRYPTO '89. Proceedings, 1989. XIII, 634 pages. 1990.

Vol. 436: B. Steinholtz, A. Sølvberg, L. Bergman (Eds.), Advanced Information Systems Engineering. Proceedings, 1990. X, 392 pages. 1990.

Vol. 437: D. Kumar (Ed.), Current Trends in SNePS – Semantic Network Processing System. Proceedings, 1989. VII, 162 pages. 1990. (Subseries LNAI).

Vol. 438: D. H. Norrie, H.-W. Six (Eds.), Computer Assisted Learning – ICCAL '90. Proceedings, 1990. VII, 467 pages. 1990.

Vol. 439: P. Gorny, M. Tauber (Eds.), Visualization in Human-Comp Interaction. Proceedings, 1988. VI, 274 pages. 1990.

Vol. 440: E.Börger, H. Kleine Büning, M. M. Richter (Eds.), CSL '89. ceedings, 1989. VI, 437 pages. 1990.

Vol. 441: T. Ito, R. H. Halstead, Jr. (Eds.), Parallel Lisp: Languages and tems. Proceedings, 1989. XII, 364 pages. 1990.

Vol. 442: M. Main, A. Melton, M. Mislove, D. Schmidt (Eds.), Mathema Foundations of Programming Semantics. Proceedings, 1989. VI, pages. 1990.

Vol. 443: M. S. Paterson (Ed.), Automata, Languages and Programm Proceedings, 1990. IX, 781 pages. 1990.

Vol. 444: S. Ramani, R. Chandrasekar, K. S. R. Anjaneyulu (Eds.), Kn ledge Based Computer Systems. Proceedings, 1989. X, 546 pages. 19 (Subseries LNAI).

Vol. 445: A. J. M. van Gasteren, On the Shape of Mathematical Argume VIII, 181 pages. 1990.

Vol. 446: L. Plümer, Termination Proofs for Logic Programs. VIII, 142 pa 1990. (Subseries LNAI).

Vol. 447: J. R. Gilbert, R. Karlsson (Eds.), SWAT 90. 2nd Scandina Workshop on Algorithm Theory. Proceedings, 1990. VI, 417 pages. 19

Vol. 449: M. E. Stickel (Ed.), 10th International Conference on Autom Deduction. Proceedings, 1990. XVI, 688 pages. 1990. (Subseries LN

Vol. 450: T. Asano, T. Ibaraki, H. Imai, T. Nishizeki (Eds.), Algorithms. ceedings, 1990. VIII, 479 pages. 1990.

Vol. 451: V. Mařík, O. Štěpánková, Z. Zdráhal (Eds.), Artificial Intellige in Higher Education. Proceedings, 1989. IX, 247 pages. 1990. (Subse LNAI).

Vol. 452: B. Rovan (Ed.), Mathematical Foundations of Computer Sci 1990. Proceedings, 1990. VIII, 544 pages. 1990.

Vol. 453: J. Seberry, J. Pieprzyk (Eds.), Advances in Cryptolo AUSCRYPT '90. Proceedings, 1990. IX, 462 pages. 1990.

Vol. 454: V. Diekert, Combinatorics on Traces. XII, 165 pages. 1990.

Vol. 455: C. A. Floudas, P. M. Pardalos, A Collection of Test Problem Constrained Global Optimization Algorithms. XIV, 180 pages. 1990.